From Web1 to Web3

From Web1 to Web3 is your definitive roadmap through the current digital revolution. Authored by Ollie Bell, Nabil Hadi, and Daniel Strode, this book offers a clear, thoughtful exploration of the internet's evolution – from its humble, static beginnings to the dynamic, decentralized future that is emerging today.

The journey begins with Web1, an era defined by a read-only landscape of information where the internet functioned primarily as a digital library. As time moved on, Web2 brought a seismic shift with its explosion of user-generated content and the rise of social media, fundamentally changing how we communicate and share. However, as centralized platforms increasingly controlled our digital interactions and data, a new need arose – a need for a system that returned control to the individual.

Enter Web3. In this new paradigm, blockchain technology, cryptocurrencies, decentralized finance (DeFi), non-fungible tokens (NFTs), and decentralized autonomous organizations (DAOs) converge to empower individuals with true digital ownership and control. Rather than relying on centralized institutions, Web3 leverages transparent, peer-to-peer networks to reimagine how we interact with the digital world.

This book provides not only a historical perspective but also practical insights for businesses and individuals alike. Through case studies featuring leading global brands and actionable guides on navigating decentralized applications (dApps), readers gain an understanding of how businesses and individuals are already using Web3 technologies to drive innovation and create value. Whether you're an entrepreneur, investor, developer, or a digital native keen to reclaim your data and identity, this book offers the knowledge you need to adapt and thrive in this rapidly evolving landscape.

Beyond the technical details, *From Web1 to Web3* explores the broader cultural and economic shifts brought about by decentralization. It examines how these changes are redefining what it means to be connected and how trust is built in a world where power is shifting from centralized authorities to individual users.

This book is a balanced and accessible guide, providing the context, analysis, and practical advice required to understand the present and future of the internet. Your journey into the evolving world of Web3 begins here.

From Web1 to Web3

Understanding the Past, Present, and Future of the Internet

Ollie Bell, Nabil Hadi, and Daniel Strode

CRC Press
Taylor & Francis Group
Boca Raton London New York

CRC Press is an imprint of the
Taylor & Francis Group, an **informa** business

Designed cover image: Ollie Bell, Nabil Hadi, and Daniel Strode

First edition published 2026
by CRC Press
2385 NW Executive Center Drive, Suite 320, Boca Raton FL 33431

and by CRC Press
4 Park Square, Milton Park, Abingdon, Oxon, OX14 4RN

CRC Press is an imprint of Taylor & Francis Group, LLC

ISBN: 978-1-041-01268-9 (hbk)
ISBN: 978-1-041-01809-4 (pbk)
ISBN: 978-1-003-61650-4 (ebk)

DOI: 10.1201/9781003616504

Typeset in Sabon
by Newgen Publishing UK

Contents

Author biographies

Ollie Bell, co-founder of Roster, brings his vast experience in education, behavioral science, and digital business transformation to the forefront of this book. With a desire to remain at the cutting edge of new thinking, Ollie's leadership is rooted in a passion for taking complex subjects and making them accessible, so that everyone can benefit in the moments that follow major breakthroughs in digital innovation. With extensive experience in education, digital innovation, and business development, Ollie works with global brands, public figures, and professional bodies, creating strategic partnerships that equip businesses to thrive in the Web3 landscape. Through Roster3.com, Ollie leads efforts to provide skills and strategies, aimed at upskilling individuals and brands within and at arm's length of the Web3 ecosystem.

Nabil Hadi, co-founder of Roster, is a tech enthusiast with a passion for innovation and a track record of success across multiple industries. He started his journey at Ingram Micro, a *Fortune 500* company, where he quickly made a name for himself in the UK IT scene – earning recognition as a rising star. From there, Nabil took the entrepreneurial leap, founding his own tech consultancy and working with global giants like McDonald's, PepsiCo, and Acer to enhance their digital experiences. But in 2016, everything changed when he discovered Bitcoin and blockchain. Fascinated by the potential of Web3, he dove deep into the space, earning certifications from the Institute of Blockchain in Singapore and Saïd Business School at Oxford University. Today, Nabil is a senior executive at a global cryptocurrency exchange and a real estate tokenization platform, where he applies his expertise in Web3 and blockchain to shape the future of digital assets.

Daniel Strode is a seasoned leader and innovator at the crossroads of strategy and culture. He is passionate about using new technology to make businesses fairer and more welcoming to everyone. As a trusted advisor to major brands, from Ferrari to AECOM and many more, Daniel's work focuses on helping organizations navigate the challenges of digital transformation. His thought leadership, especially in Artificial Intelligence and corporate culture, drives his commitment to empowering businesses to not only adopt Web3 technologies but also to build people-first cultures that embrace diversity and community-driven innovation. Daniel's first book, *The Culture Advantage* (Kogan Page, 2022), is an exploration of the role corporate culture plays in driving sustainable growth in the face of constant change and his second, *The Innovator's Edge* (Independently Published, 2024), expands upon his work identifying hundreds of insightful tips and tricks to drive innovation across an organization.

Acknowledgments

The book you are holding, whilst written by Ollie, Nabil and Daniel, is mostly the product of Roster's incredible team of advisors, all of whom have been integral to shaping its content. This team represents some of the brightest minds in the Web3 space, each bringing their expertise to the table to ensure this book delivers the most accurate, timely, and actionable information – it is a real rarity to have been able to call upon such a breadth of knowledge.

The core Roster team and other co-founders are really a crack set of experts. The team is composed of Matt Warner and Tom Downing. Tom brings over a decade of expertise in the UK creative tech agency space, having advised *Fortune 500* companies across multiple industries, with a passion for innovating consumer experiences through emerging technologies. Matt is a multifaceted professional with a diverse background spanning athletics, media, and creative design, where he has worked with top sports and fitness personalities while leading visual design initiatives for major brands. While each of them have all made unbelievable contributions to the book and the creation of the Roster3.com digital-certified MBA and Masterclasses – this book is a continuation of their effort and commitment. Alongside is the wonderful *Roster* of advisors, who each have contributed their time and expertise to create world class educational content on their specific topics that is truly pushing individuals and businesses forward, embracing Web3. As such, special thanks goes to:

- **Dr. Niaz Chowdhury**, an educator and researcher with a deep focus on blockchain and cryptocurrencies and author of *Inside Blockchain, Bitcoin and Cryptocurrencies* (Auerbach, 2019), has provided the critical understanding of the technical underpinnings of Web3 in this book.
- **Vineeth Bhuvanagiri**, CEO of EMURGO Fintech, lends his expertise in blockchain and financial technologies, particularly in the context of DeFi. His views really break down how DeFi is revolutionizing the financial world.

- **Anndy Lian** is an all-rounded business strategist in Asia. He has provided advisory across a variety of industries for local, international, and public-listed companies and governments. He is an early blockchain adopter and experienced serial entrepreneur, book author, investors, board member, and keynote speaker.
- **Aleksa Mil**, founder of Daoalexa, offers Roster insights on governance, token design, and community management that are essential for building sustainable Web3 organizations.
- **Nova Lorraine** is a psychologist turned award-winning fashion designer and futurist. As the founder of *House of Nova*, a digitally native culture brand, and author of *The Jockey on the Horse* (Independently Published, 2024), she pioneers storytelling and conscious innovation at the intersection of fashion, Web3 and AI.
- **Gareth Malna** is a regulatory lawyer, partner at Gunner Cooke and the co-founder of Englebert, which has allowed some of the largest crypto asset businesses in the world to market compliantly in the United Kingdom.

A special thanks is sent to the Head of Research at Roster, **Nahin Talhah**, who is a mathematician studying at King's College London, working on combining his academic background with real-world blockchain applications. While deeply interested in DeFi, particularly on-chain liquidity dynamics and crypto-based derivatives, his focus lies in privacy-enhancing technologies.

The collaborative effort of this talented group has resulted in a book that not only covers the essential aspects of Web3 but also serves as a roadmap for businesses looking to thrive in the rapidly evolving decentralized digital landscape. With their combined expertise, this book will empower you to understand Web3, embrace its challenges, and create real value. As we say:

> "Welcome to Web3. Your adventure begins now."

Glossary

Automated Market Makers (AMMs): Use smart contracts to automatically set the price of an asset based on the supply and demand of various cryptocurrencies. Most take the shape of an $x*y = k$ curve with supply and demand acting inversely proportional to each other, **differing from** traditional order book methods.

Blockchain: A distributed ledger technology (DLT) that records and stores data across a decentralized network of computers.

Blockchain-Based Smart Contracts: Automatically execute agreements without the need for an intermediary.

Consensus Mechanism: A process where multiple participants (nodes) agree on the validity of transactions on a shared network (blockchain).

Cryptocurrency: A digital or virtual currency secured by cryptography.

Custodial Wallet: A wallet where a third party holds the private keys on your behalf.

dApps (Decentralized Applications): Applications that run on blockchain networks instead of traditional centralized servers.

Data Ownership: The concept that users own and control their own data in decentralized networks.

Decentralized Autonomous Organizations (DAOs): Organizations that are run communally by smart contracts and governed by token holders rather than a central authority.

Decentralized Exchange (DEX): An exchange built on a blockchain that removes traditional financial intermediaries.

DeFi (Decentralized Finance): A suite of financial services that operate on blockchain networks, eliminating intermediaries.

Digital Identity: Representation of one's identity in the digital space, controlled by the individual in Web3.

Digital Ownership: The concept that digital assets are controlled by the user, not a platform, in Web3.

Digital Wallet: A secure application that allows individuals to store, manage, and access their digital assets.

Distributed Ledger Technology (DLT): A technology that records and stores data across a decentralized network of computers.

E-money: Electronically stored monetary value that can be used to make payments.

Ethereum: A widely used smart contract blockchain platform for Web3 applications.

Ethereum Virtual Machine (EVM): A digitized computer that is shared throughout the globe to run and execute all code on the Ethereum network.

Fungible: Interchangeable with one another.

Governance Token: A digital asset that provides holders a voice in the decision-making processes of a system.

Immutability: Data written into a block that cannot be altered or erased.

Interoperability: The ability for digital assets, identity, and data to move across different platforms.

Layer 2 Solutions: Solutions built on top of Layer 1 blockchains to optimize the performance of the underlying Layer 1.

Liquidity Pools: Where users can stake their assets to provide liquidity for trades, earning fees in return.

Metadata: Key information about an NFT, such as its title, description, and media files.

Nodes: Computers on a blockchain network that validate transactions and maintain the network.

Non-Custodial Wallet: A wallet that gives you full control over your private keys.

Non-Fungible Tokens (NFTs): Unique digital assets that represent ownership of a specific item.

Open Source: A system where anyone can participate in the development and governance of a platform.

Peer-to-Peer Marketplace: A marketplace where individuals can trade directly with one another.

Peer-to-Peer Network: A network where all participants have an equal opportunity to directly validate and record transactions without the need for a central authority's approval.

Private Keys: Cryptographic keys that prove your ownership over a digital wallet.

Proof-of-Stake (PoS): A consensus mechanism where validators lock up cryptocurrency as collateral and malicious activity typically results in the loss of collateral.

Public Key: A cryptographic code (similar to an email address) that is used to receive digital assets or data to a particular digital wallet.

Quadratic Voting: A system where one token does not equal one vote, instead votes scale quadratically causing large holders to have proportionally lower voting power relative to their token amount.

Read-Write-Own Internet: The concept of Web3 where users are empowered to fully own and control their digital lives.

Regulatory Uncertainty: The regulatory landscape for Web3 technologies is still evolving and not always clear.

Royalty Feature: A mechanism in NFTs that ensures creators receive a percentage of each subsequent sale.

Scalability: The ability of a blockchain to handle a large number of transactions.

Security Audit: A thorough review of the code to ensure there are no vulnerabilities.

Self-Custody: The concept that users control their own digital assets without reliance on a third party.

Self-Sovereign Identity (SSI): An identity system based on blockchain; individuals create their digital identity on the blockchain.

Smart Contracts: Self-executing contracts with the terms of the agreement directly written into lines of code.

Staking: Agreeing with a network or dApp your assets will not be sold over a period of time for the promise of additional assets after the allotted time period ends.

Token Economies: (Or Tokenomics) New business models where digital tokens are used to incentivize certain behaviors.

Traceable: Meaning anyone in the network can view the transaction's history.

Transaction Fees: Costs associated with transactions on a blockchain network.

Transparent: Meaning transaction's details are viewable to anyone on the network.

True Ownership: Digital ownership is transferred directly to the individual, thanks to blockchain technology.

Trustless Nature: A characteristic of blockchain transactions that eliminates the need for trust in intermediaries through consensus mechanisms.

Validators: Individuals who check transactions to ensure they meet protocol rules (in proof-of-stake systems).

Virtual Real Estate: Digital land that can be bought, sold, and developed using NFTs.

Virtual Reality (VR): A technology that allows users to fully immerse themselves in virtual environments.

Web1: The dawn of the internet, a static, read-only web.

Web2: The internet that evolved into a more dynamic and interactive space.

Web2.5 Wallets: Wallets that combine aspects of traditional Web2 and Web3 wallets.

Web3: The next leap in the evolution of the internet that moves beyond the centralized data model of Web2 and embraces decentralization.

Weighted Voting: Allows for greater influence for those who have contributed more to the DAO.

Yield Farming: Lending your crypto to DeFi protocols or liquidity pools in exchange for rewards.

Zero-Knowledge Proofs (ZKPs):   Allow for the verification of transactions without revealing the transaction data itself.

Disclaimer

This book is written for informational purposes only and does not constitute financial advice. The content reflects the best knowledge and understanding of the contributors at the time of publication. The rapidly evolving nature of technology and the regulatory environment surrounding the topics discussed means that the information provided may not be accurate or up to date at the time you are reading it. Readers should conduct their own research and consult with qualified professionals before making any financial or business decisions related to the content covered in this book.

Neither the publisher nor the authors can be held liable for any actions, decisions, or consequences arising from the use of this book's information. The technologies, platforms, and regulations described herein are subject to constant change and variation, and it is essential to stay informed of current developments. The publisher and authors disclaim any liability for any loss or damage that may occur as a result of using this book's content.

Chapter 1

The Web3 evolution

Navigating a new digital frontier

Picture this: It's a crisp morning, and you're sipping your coffee, scrolling through your decentralized social media feed. The posts you see aren't determined by some mysterious algorithm designed to keep you scrolling endlessly. Instead, they're curated by you and your community, reflecting your genuine interests and connections. Your data isn't being harvested and sold to the highest bidder. In fact, you're earning cryptocurrency tokens just for participating in the network.

Welcome to the world of Web3, where the digital landscape has been transformed from a corporate-controlled oligarchy to a user-empowered democracy. This isn't some far-off sci-fi fantasy. It's the reality unfolding right now. We're standing at the precipice of a digital revolution set to reshape not just how we interact online, but how we conceptualize ownership, value, and even identity in the digital age. And you are about to become a part of it.

For decades, we've lived in a digital world dominated by tech giants. These behemoths have shaped our online experiences, controlled our data, and essentially acted as the gatekeepers of the internet. They've brought us incredible innovations, no doubt, but at what cost? Our privacy has been eroded, our attention commodified, and our digital lives have become products to be bought and sold.

But the tides are turning. Web3 is here to flip the script and put the power back where it belongs – in your hands. At its core, Web3 is built on the foundation of blockchain technology, cryptocurrencies, and decentralized applications (dApps). These aren't just buzzwords; they're the building blocks of a new digital democracy. As recent research shows, the world's top brands are already adopting Web3 strategies, preparing for its mainstream adoption. Web3 is not just a set of technologies; it's an entire rethinking of digital interaction.

DOI: 10.1201/9781003616504-1

1

THE FOUNDATIONS OF A DIGITAL REVOLUTION

Imagine a world where you truly own your digital assets. Where your online identity isn't tied to a Facebook or Google account but is a sovereign entity that you control. A world where content creators can connect directly with their audience without intermediaries taking a cut. Where your financial transactions aren't monitored and controlled by banks, but are secure, transparent, and entirely under your control.

This is the promise of Web3, and it's not just theoretical. It's happening right now, all around us. Decentralized finance (DeFi) platforms are already challenging traditional banking systems, offering financial services without the need for intermediaries. Non-fungible tokens (NFTs) are revolutionizing how we think about digital ownership and creativity. Decentralized autonomous organizations (DAOs) are reimagining how we structure and govern organizations. And the Metaverse is opening up entirely new realms for social interaction, commerce, and creativity. Leading brands, from Nike to Gucci, are exploring the Metaverse and Web3 technologies, signaling a shift in the way businesses engage with consumers in digital spaces.

But let's be clear: the road ahead isn't without its challenges. Web3 is still in its infancy, and like any transformative technology, it faces hurdles. There are technical challenges to overcome, regulatory landscapes to navigate, and mindsets to shift. Yet, these challenges are precisely what make this moment so exciting. We're not just passive consumers of a finished product; we're active participants in shaping this new digital frontier. Every developer who builds a dApp, every artist who mints an NFT, and every user who chooses a decentralized platform over a centralized one – they're all contributing to the evolution of Web3.

So, how do you become a part of this revolution? It starts with understanding. Understanding the technologies that underpin Web3, the principles that drive it, and the possibilities it opens up. We have interviewed, researched, and worked with hundreds of leading brands, and we're seeing a 'test and learn' approach to Web3 adoption. Major players, such as Intel and Goldman Sachs, are experimenting with blockchain and NFTs, realizing that these technologies are key to future-proofing their businesses. They're learning from these initial ventures, laying the groundwork for deeper integration in the years to come. That's where this book comes in. We're going to take a deep dive into the world of Web3, exploring its foundational technologies, its current applications, and its future potential.

BEYOND TECHNOLOGY: A NEW DIGITAL PARADIGM

We'll start by demystifying blockchain technology. You've probably heard of Bitcoin and other cryptocurrencies, but blockchain is so much more than just digital money. It's a decentralized ledger that enables trust and transparency in digital transactions without the need for intermediaries. We'll explore how blockchain works, its different types, and how it's being used beyond cryptocurrencies.

Then, we'll look at the world of decentralized finance, or DeFi. DeFi is reimagining financial services for the digital age, offering everything from lending and borrowing to insurance and trading, all without the need for traditional banks or financial institutions. We'll look at how DeFi works, its potential to democratize finance, and the challenges it faces.

Next, we'll explore the fascinating world of non-fungible tokens, or NFTs. These unique digital assets are revolutionizing how we think about ownership in the digital realm. From digital art to virtual real estate, NFTs are opening up new possibilities for creators and collectors alike. We'll dive into how NFTs work, their impact on various industries, and their potential future applications.

We'll also take a look at decentralized autonomous organizations, or DAOs. These novel organizational structures are challenging traditional notions of corporate governance. Imagine an organization where decisions are made not by a board of directors, but by a community of token holders. We'll explore how DAOs work, their potential to revolutionize organizational structures, and the challenges they face in implementation.

And of course, we can't talk about Web3 without discussing the Metaverse. This concept of a persistent, shared, 3D virtual space is capturing imaginations worldwide. We'll explore what the Metaverse is, how it relates to Web3, and its potential to reshape how we work, play, and interact online.

But understanding the technology is only half the battle. The real power of Web3 lies in its application. That's why you'll be peppered throughout this book with real-world case studies and examples. We'll look at how forward-thinking businesses, like Gucci, Adidas, and Nike, are already leveraging Web3 technologies to innovate and create value. From startups disrupting traditional industries to established companies pivoting to embrace the Web3 future, these stories will provide practical insights and inspiration for your own Web3 journey. These stories and more will illustrate the real-world impact and potential of Web3 technologies.

But this book isn't just about understanding Web3 – it's about empowering you to be a part of it. That's why each chapter will include practical guides and actionable steps. We'll walk you through setting up your first cryptocurrency wallet, participating in a DeFi protocol, minting an NFT, and more. By the time you finish this book, you won't just understand Web3 – you'll be an active member.

We'll also address the challenges and potential pitfalls of Web3. Like any emerging technology, Web3 faces its share of blockers. There are technical challenges to overcome, such as scalability issues and energy consumption concerns. There are user experience challenges as many Web3 applications are still complex and intimidating for the average user. In addition there are regulatory challenges, as governments and regulatory bodies grapple with how to approach this new decentralized landscape. We'll tackle these issues head-on, exploring potential solutions and ongoing developments.

And finally, we will close with a full look at how artificial intelligence (AI) is starting to change everything as we know it, and how the partnership between AI and Web3 is potent for both businesses and individuals.

BUILDING THE DECENTRALIZED FUTURE

Throughout our journey, we'll look to the future. What does the widespread adoption of Web3 technologies mean for society? How might it reshape our economies, our governance structures, even our concept of national borders? We'll explore these big questions, not with the aim of predicting the future, but of preparing you to be an active participant in shaping it. Because that's what this book is really about. It's not just a guide to understanding Web3 – it's a call to action. The future of the internet is being written right now, and you have the opportunity to be one of its authors. Whether you're a developer looking to build the next groundbreaking dApp, an entrepreneur seeking to disrupt your industry with blockchain technology, or simply an individual eager to take control of your digital life or change the path of your company, Web3 offers unprecedented opportunities.

The shift from Web2 to Web3 isn't just a technological evolution – it's a paradigm shift in how we interact with the digital world. It's a move from centralization to decentralization, from corporate control to user empowerment, from closed systems to open protocols. It's a chance to rectify the mistakes of the past and build a more equitable, transparent, and user-centric digital future.

But this future won't build itself. It requires visionaries, builders, and pioneers. It needs people who are willing to challenge the status quo, to imagine new possibilities, and to put in the work to make those possibilities a reality. In short, it needs people like you. As we stand on the brink of this new digital frontier, the question isn't whether Web3 will change the world – it's already doing that. The question is: what role will you play in shaping this change? Will you be a passive observer, watching from the sidelines as others define the future of the internet? Or will you be an active participant, leveraging these new technologies to create value, solve problems, and push the boundaries of what's possible?

The choice is yours, and the time to make that choice is now. Web3 isn't some far-off future – it's happening right now, all around us. Every day, new Web3 projects are launching, new use cases are being discovered, and new possibilities are being unlocked. Established companies are more than dipping their toes into the water; they are truly embracing this technology and bringing products and services to the mass consumer market already. The opportunity to be at the forefront of this revolution is here, right now.

So, are you ready to take the plunge? To dive into the world of Web3 and emerge as a builder of the decentralized future? If so, then let's get started. The pages that follow are your roadmap to understanding, participating in, and shaping the Web3 revolution. Remember, every great journey begins with a single step. Your first step into the world of Web3 starts here, with this book. So take a deep breath, turn the page, and let's embark on this exciting journey together. The future of the internet is decentralized, and it's waiting for you to help shape it.

> "Welcome to Web3. Your adventure begins now."

Chapter 2

Foundations of Web3

Concepts and core principles

As technology shifts and evolves at breakneck speeds we are seeing a transformation unfold in front of our eyes. In fact, we often feel that if we blink our eyes for even a second we have missed a whole revolution in that instance, at least in terms of technology.

At Roster we feel that every day in our business, and the key transformation unfolding right now is **Web3** – which encompasses multiple technologies.

But what exactly is Web3, and how did we get here? To understand this new era of digital innovation, we must first step back and explore the previous versions of the web: **Web1** and **Web2**. These earlier versions laid the groundwork for the decentralized, self-custody future that Web3 promises. However, to truly comprehend Web3's potential, we need to explore its core features, implications, and the ways in which it will revolutionize how we interact with the digital world.

THE BIRTH OF THE INFORMATION ECONOMY: WEB I

Web1, also referred to as the Information Economy, represents the dawn of the internet as we know it. Emerging in the early 1990s, Web1 was a static, read-only web where users primarily consumed content. The internet was a vast repository of information, but interaction was non-existent – users were merely visitors to websites that presented data in the form of text-heavy, unchanging pages. The very concept of user-generated content (UGC) was a distant thought, with everything controlled by webmasters and institutions.

For most of the early 1990s, the internet was dominated by academic institutions and government agencies, offering a limited set of use cases primarily focused on research and knowledge sharing. Websites were **simple HTML documents, static in nature,** and primarily served as digital brochures or information pamphlets. There were no comments, likes, or

DOI: 10.1201/9781003616504-2

shares; no features for users to engage with the content. The experience was limited to consuming information rather than creating it.

An example of how rudimentary the early web was can be found by looking at eBay's website in 1997. What was once considered cutting edge, upon reflection feels clunky, bare-bones, and basic. Despite its simplicity, Web1 was a crucial turning point. It laid the foundation for the modern internet and sparked a shift in how people could access and share knowledge.

Web1 was characterized by centralized control over the flow of information. A select few, typically large corporations and institutions, controlled what content was published, and they dictated the means of distribution. It was the first time information could be shared at global scale, though still within a highly restricted ecosystem. The tools were clumsy, but the concept of a global web of knowledge was born, paving the way for the more dynamic internet that followed.

Although Web1 was limited in interactivity, its significance cannot be overstated – it set the stage for the internet's rapid growth. It marked the beginning of a new world where information could be shared instantly, around the clock, and across the globe. It was only a matter of time before users would demand more interactivity and personalization, thus laying the groundwork for Web2.

THE RISE OF USER-GENERATED CONTENT: WEB2

As technology progressed and internet infrastructure improved, Web1's limitations became more apparent. In the early 2000s, the internet began to evolve into a far more dynamic and interactive space, giving rise to Web2, or the Creator Economy.

The introduction of Web2 marked a shift from passive consumption to active creation. This era heralded the proliferation of UGC, social media, and the rise of platforms such as YouTube, Facebook, and X (formerly Twitter).

The defining characteristic of Web2 was the concept of a 'read-and-write' internet, where users transitioned from passive content consumers to active participants. This transformation was propelled by several technological innovations. Faster internet speeds, the advent of broadband, the widespread adoption of smartphones, and the availability of high-bandwidth networks all contributed to this newfound interactivity. As data speeds accelerated, Web2 unlocked a world where users could not only consume information but could also contribute and engage on a global stage.

Platforms like Facebook, Instagram, and X empowered users to create profiles, upload photos, and interact with others in ways that were unimaginable during Web1. Social networks emerged, fostering communities around common interests. This democratization of content creation allowed anyone with an internet connection to participate in the creation and dissemination of content.

Now, people weren't just consumers; they were prosumers – individuals who created and shared content that had the potential to reach millions of people worldwide.

Web2 also ushered in the era of the sharing economy and digital marketplaces. Sites like eBay, Amazon, and Etsy allowed users to not only buy and sell products but also to curate their digital storefronts. The very nature of the internet had evolved from a digital library into a dynamic marketplace.

However, with this new wave of interaction came a new set of challenges, most notably the increasing dominance of big tech. In Web2, platforms like Google, Facebook, and Amazon gained enormous power.

These companies controlled the flow of information, leveraged vast amounts of user data, and profited from the personalized services they provided. While the user experience improved, and interactivity soared, the price for this convenience was steep: privacy was compromised.

Web2 also created a data-driven economy, where UGC became a commodity. In this system, platforms monetized user activity, often without users being fully aware of the scope of data being harvested. Personal data became the currency of the web, and individuals traded their privacy for access to services. In this new economy, the internet was no longer a tool for simple access to information – it became a critical component of every aspect of modern life, from shopping to socializing to entertainment.

Yet, as we began to see the limits of Web2 – centralized control, surveillance capitalism, and privacy concerns – it became evident that the next phase of the internet would need to offer a new vision: one of self-ownership, privacy, and decentralization. Enter Web3.

> *"The original idea of the web was that it should be a collaborative space where you can communicate through sharing information."*
> Tim Berners-Lee (Inventor of the World Wide Web)

WEB3: THE SELF-CUSTODY AND OWNERSHIP ECONOMY

Web3 represents the next leap in the evolution of the internet. It moves beyond the centralized data model of Web2 and embraces decentralization, giving power and control back to the individual. Web3 builds on the foundation laid by Web1 and Web2 but introduces an entirely new framework for interacting with digital assets, data, and identity. At its core, Web3 is about removing intermediaries and empowering individuals to control their digital lives.

The central technological innovation behind Web3 is blockchain. Blockchain technology enables secure, transparent, and decentralized transactions without the need for traditional intermediaries such as banks, tech giants, or government entities. Initially known for enabling cryptocurrencies like Bitcoin and Ethereum, blockchain's true potential lies in its ability to allow for secure, verifiable transactions across a wide range of applications. This technology allows assets – whether digital tokens, personal data, or identities – to be transferred directly between individuals, removing the need for a central authority.

One of the most revolutionary aspects of Web3 is the concept of **self-custody** and **true ownership**. In Web2 and across the creator economy, when we create or purchase something online – be it a photo, a song, or even a social media post – ownership typically resides with the platform. For example, when you upload a photo to Instagram, Instagram retains ownership rights to that content. In Web3, however, digital ownership is transferred directly to the individual, thanks to blockchain technology. This means that assets like artwork, cryptocurrencies, and even your social media profiles are truly yours, and you can control, sell, or transfer them as you wish.

A key tool in Web3 is the **digital wallet**, which serves as the gateway to your assets in the decentralized web. Your wallet stores private keys, which are akin to passwords that protect your digital identity and assets. With this tool, you can manage and control your digital assets – whether it's cryptocurrency, non-fungible tokens (NFTs), or personal data – without relying on a centralized platform. This represents a major shift from Web2, where users had to trust platforms with their data.

In addition to digital wallets, Web3 also introduces **decentralized applications (dApps)**. These are applications that run on blockchain networks instead of traditional centralized servers. dApps are designed to be transparent, secure, and resistant to censorship. They enable users to interact with the digital world without the need for centralized intermediaries, offering the possibility of a more democratic and open web.

Whether it's decentralized finance (DeFi), decentralized social networks, or marketplaces, Web3 dApps empower users to control their own online experiences and transactions. All without the oversight or influence of powerful centralized corporations, and we will be taking a much closer look at those in the coming chapters.

The transition to Web3 represents a profound shift in how we think about digital ownership, identity, and control. It's about creating a **read-write-own** internet, where users are no longer bound by centralized entities but are empowered to fully own and control their digital lives.

As Web3 continues to develop, its implications for industries ranging from finance and healthcare to media and entertainment are vast. This new paradigm is not just about technology – it's about **fundamentally rethinking how** value is created, shared, and controlled on the internet. And with Web3, the next era of the internet promises to be one where the user, not the platform, is in control.

REAL-WORLD EXAMPLES OF WEB3 IN ACTION

To understand the true potential of Web3, it helps to look at some **real-world examples** of how this technology is already being used by businesses and individuals around the world. These case studies illustrate the wide range of applications for Web3, from digital ownership to DeFi.

1. **Nike and digital ownership: Reimagining the sneaker culture**
    Nike, a global titan in the sportswear industry, has always been at the forefront of innovation. As Web3 technologies began to gain traction, Nike quickly recognized the immense potential to not just sell physical products but to offer a new form of ownership that extends into the virtual world. Their foray into Web3, particularly through the use of **NFTs**, represents a bold step into the future of both fashion and digital interaction.

    In this new Web3 model, Nike has created a seamless blend of the real and virtual worlds by offering **digital twins** of their iconic sneakers. These are unique digital representations of physical products that exist on the blockchain as NFTs. When a customer buys a physical pair of Nike sneakers, they also get the option to purchase the digital twin – an exact replica of their shoes in virtual form. This digital asset can then be worn in virtual environments like video games, metaverses, or other online spaces that support NFTs.

    But Nike's integration with Web3 doesn't stop at mere digital ownership. Their NFTs open up exciting new opportunities for customers to engage with the brand and each other. For instance, Nike has partnered and tested solutions with platforms like **RTFKT Studios**, known for its work in the virtual fashion space, to create limited-edition sneakers and apparel exclusively for the metaverse. These collaborations introduce virtual wearables, which cannot be copied or replicated, ensuring that owners have true digital ownership of their assets – just as they would with physical sneakers. The idea is not only to expand into virtual spaces but also to offer customers the ability to trade, collect, and sell these digital items in secondary markets, further enriching their connection with the brand. Although the RTFKT test is now finished, there will be more.

    The use of blockchain technology in Nike's NFT strategy allows for verifiable ownership and scarcity – traits that are essential in the world of collectibles. By leveraging **Ethereum** and other blockchain protocols, Nike ensures that their digital products are unique, traceable, and secure, giving consumers a sense of pride in owning something that is both limited in edition and backed by blockchain-based proof of authenticity.

This shift toward digital ownership represents a future where consumers own a portion of their brand interactions, transcending the traditional model of simply purchasing physical goods.

2. **Audius and the decentralized music revolution**

Audius, a decentralized music streaming platform, is a groundbreaking example of Web3 technology reshaping an entire industry. In the world of Web2, artists have historically been at the mercy of centralized platforms like **Spotify** and **Apple Music**, which dominate the music streaming market.

While these platforms provide massive reach, they also impose strict control over revenue-sharing models, typically taking a large commission and leaving artists with only a fraction of the profits from their work – there is hardly a day that goes by without the news reporting about an artist not being paid, or others complaining over the royalties.

Audius flips this traditional model on its head by offering a **decentralized** alternative that empowers musicians. Built on the **Ethereum blockchain**, Audius enables artists to upload their music directly onto the platform and retain full ownership and control over their content. By removing the middlemen – such as record labels and streaming platforms – Audius allows artists to earn a larger share of the revenue generated from their music.

The platform operates on **blockchain-based smart contracts**, which automatically execute agreements between artists and their fans without the need for an intermediary. These contracts allow for a transparent and fair distribution of revenue. Artists receive payment in the form of **Audius' native token, AUDIO**, which is directly transferred from fans who listen to or support their music. This creates a truly peer-to-peer ecosystem where the artist and their audience have a direct financial relationship, bypassing the profit-centric control of corporate intermediaries.

Furthermore, Audius incorporates the concept of **tokenization** to allow fans to become more than just passive listeners. Through the AUDIO token, users can vote on important decisions, participate in governance, and even gain exclusive access to content from their favorite artists. This transforms the fan experience into something much more interactive and participatory. In essence, Audius represents a Web3-powered music ecosystem that gives creators control, transparency, and a direct link to their supporters – fundamentally reshaping the way music is monetized and consumed.

In doing so, Audius demonstrates how Web3 is not only empowering individual creators but also contributing to the broader movement of

DeFi, where intermediaries are replaced with more open, transparent, and user-centered models.

3. **Uniswap and the DeFi movement**

    Uniswap, a decentralized exchange (DEX) built on the **Ethereum blockchain**, is one of the most prominent examples of how Web3 is revolutionizing the financial landscape. Traditional financial systems rely on centralized institutions – like banks, stock exchanges, and brokers – to facilitate transactions, set prices, and ensure liquidity. These inter-mediaries take a cut of every transaction and, in many cases, impose barriers to entry for people around the world.

    Uniswap, however, removes these intermediaries entirely. Instead of using a traditional order book to match buyers and sellers, Uniswap operates using **automated market makers (AMMs)**. These AMMs use smart contracts to automatically set the price of an asset based on the supply and demand of various cryptocurrencies. (Most but not all AMM's take the shape of an $x*y = k$ curve with supply and demand acting inversely proportional to each other, differing from traditional order book methods where bid-and ask-orders are matched to complete a transaction.) In essence, Uniswap functions as a **peer-to-peer** marketplace where individ-uals can trade directly with one another, and all transactions are processed through secure, transparent, and immutable blockchain networks.

    Uniswap has democratized access to financial markets by enabling anyone with an internet connection to trade cryptocurrencies without needing approval from a centralized authority. This model is particularly empowering for individuals in countries with underdeveloped financial infrastructures or places where access to traditional banking services is limited.

    The decentralized nature of Uniswap means that no one central entity controls the flow of transactions, ensuring that the exchange is resistant to censorship and manipulation. By utilizing **liquidity pools**, where users can stake their assets to provide liquidity for trades, Uniswap incentivizes users to participate and earn fees in return. Liquidity providers – people who deposit their tokens into these pools – are compensated with a portion of the trading fees.

    In terms of scalability and future possibilities, Uniswap also exemplifies the **power of interoperability**. The platform enables the integration of various tokenized assets from across different blockchain networks, making it a highly versatile tool for a broad range of crypto-related activ-ities. It also serves as a key building block for the broader **DeFi** eco-system, where users can not only trade assets but also lend, borrow, and earn interest on their holdings, all without needing to trust traditional financial institutions.

Uniswap's **open-source** nature further underscores the principles of Web3, as anyone can participate in the development and governance of the platform, ensuring that control remains decentralized. This peer-to-peer infrastructure is part of the broader trend of moving away from centralized financial systems toward more inclusive, transparent, and open alternatives.

4. **Hivemapper: How Web3 incentives create superior products**
    Hivemapper is a decentralized mapping platform that enables individuals to contribute and update global maps through the use of dashcams. The platform allows users, called 'mappers,' to collect road and map data as they drive, achieving a superior mapping tool as data is frequently updated compared to popular alternatives that rely on legacy systems.

    Hivemapper is powered by blockchain technology on Solana, which ensures transparency and integrity within its business. As Hivemapper is a blockchain-native company it prefers novel reward mechanics such as cryptocurrency-based incentive models to provide monetary stimulants to 'mappers' ensuring efficient data generation.

    By rewarding 'mappers' for their contributions, Hivemapper taps into a global community of users who are motivated to keep maps current and accurate.

The incentive structure leads to a better product because:

1. **Constant updates**: Since mappers are motivated to collect data regularly, the maps are constantly updated, ensuring more accurate and up-to-date information.
2. **Broad coverage**: With users around the world contributing, the platform can gather data from a wide range of locations, including rural or less-documented areas that are often neglected by traditional mapping services. In this case 18 million kilometers of unique mapping data.
3. **Community engagement**: The use of blockchain for rewards fosters a sense of ownership and involvement in the mapping process, creating a more engaged and dedicated community of contributors.
4. **Efficiency**: By utilizing a decentralized network, Hivemapper leverages the power of the 'hive' to gather data more quickly and efficiently than traditional centralized systems, which is why Hivemapper is the fastest growing mapping company in history, with 30% of the world covered to date.

Ultimately, the incentive-based model leads to higher-quality, more reliable maps that are always evolving to reflect real-world changes, benefiting both the mappers and the users of the maps.

THE BUILDING BLOCKS OF WEB3: BLOCKCHAIN AND DECENTRALIZATION

To fully grasp the potential of **Web3**, it's essential to first understand the technologies that power it. While Web3 introduces a revolutionary shift in how we interact with the internet, its foundation rests on blockchain technology. Blockchain is often associated with cryptocurrencies like Bitcoin and Ethereum, but its uses extend far beyond digital currencies. Let's break down **what blockchain is, how it functions,** and **why it is the backbone of Web3.**

WHAT IS BLOCKCHAIN?

At its core, a **blockchain** is a **distributed ledger technology** (DLT) that records and stores data across a decentralized network of computers, known as **nodes**. This data is stored in 'blocks,' which are linked together in a chronological order to form a continuous chain. These blocks are **immutable**, meaning once data is written into a block, it cannot be altered or erased. This property ensures that data remains secure, transparent, and resistant to tampering or fraud.

Imagine a digital **ledger** that everyone can see and verify, but no one can change once it's written. Blockchain's most distinguishing feature is its decentralized nature. In traditional systems, such as banks or centralized databases, a single authority controls and validates transactions. In contrast, **blockchain** operates without such intermediaries. Instead, it relies on a **consensus mechanism,** where multiple participants (nodes) validate and agree on the accuracy of transactions.

> *"Blockchain is like a great big notebook that everyone can read freely and for free, and write in, but one which is impossible to destroy and whose contents no one can erase."*
> Jean-Paul Delahaye (computer scientist and mathematician)

For example, when you initiate a cryptocurrency transaction, such as sending Bitcoin to someone, that transaction is broadcast to the network.

Miners (in proof-of-work systems such as Bitcoin) or validators (in proof-of-stake systems) check the transaction to ensure that it meets the protocol's rules. Once validated, the transaction is permanently recorded in a new block, which is then linked to the previous block. This process is **transparent** and **traceable**, meaning anyone in the network can view the transaction's history.

Blockchain technology fundamentally changes how we trust systems. **Trust** in traditional systems comes from intermediaries like banks or corporations. In Web3, trust comes from the transparency and consensus-driven mechanisms inherent in blockchain. There's no need for a middleman – transactions are verified and recorded by the collective network, ensuring that everyone can trust the data without relying on a central authority.

HOW BLOCKCHAIN ENABLES WEB3

Blockchain is the **driving force** behind the decentralized nature of **Web3**. Unlike traditional applications (or **dApps**) hosted on centralized servers owned by one company, **dApps** run on blockchain networks. This means that no single entity controls these applications, and they operate autonomously based on the rules encoded in their **smart contracts**.

A **smart contract** is a self-executing contract with the terms of the agreement directly written into lines of code. It automatically executes actions based on predefined conditions without needing a trusted third party. In the physical world we may liken this to a vending machine in the sense that you put a certain, predefined, amount of money in and a certain, preknown, drink or snack will fall out of the machine, as seen in Figure 2.1. On the blockchain, for example, a smart contract might trigger the transfer of cryptocurrency once a specific action is completed, such as the delivery of goods in an online marketplace. Because these contracts run on blockchain, they are **immutable** and transparent, ensuring all parties involved can see the contract's terms and are assured that the agreed-upon actions will take place.

Web3 promises to democratize the internet. Blockchain enables users to **own** their data, **control** their digital assets, and interact with others in a way that's autonomous, secure, and free from centralized control. Through blockchain, Web3 is setting the stage for new business models, such as token economies, **digital ownership**, and **DeFi**, where transactions can happen directly between users without relying on intermediaries.

THE IMPORTANCE OF SELF-CUSTODY AND OWNERSHIP IN WEB3

One of the most transformative aspects of Web3 is its focus on **self-custody** and **digital ownership**. In the traditional Web2 world, platforms such as **Facebook**, **Google**, and **Amazon** hold complete control over the data we generate. For instance, when you upload a photo to Facebook or tweet on X, you don't own those assets. Instead, the platform retains ownership rights and can even remove or alter the content.

In Web3, ownership shifts to the individual. Blockchain technology ensures that digital assets – whether it's cryptocurrency, NFTs, or other forms of data – are controlled by the user. The user becomes the **true owner** of their assets, without relying on a third party. This is where **digital wallets** come into play.

HOW DIGITAL WALLETS WORK IN WEB3

A **digital wallet** is a secure application that allows individuals to store, manage, and access their digital assets, such as **cryptocurrencies** and **NFTs**. These wallets generate **private keys**, which are cryptographic keys that prove your ownership of an asset on the blockchain. Think of the private

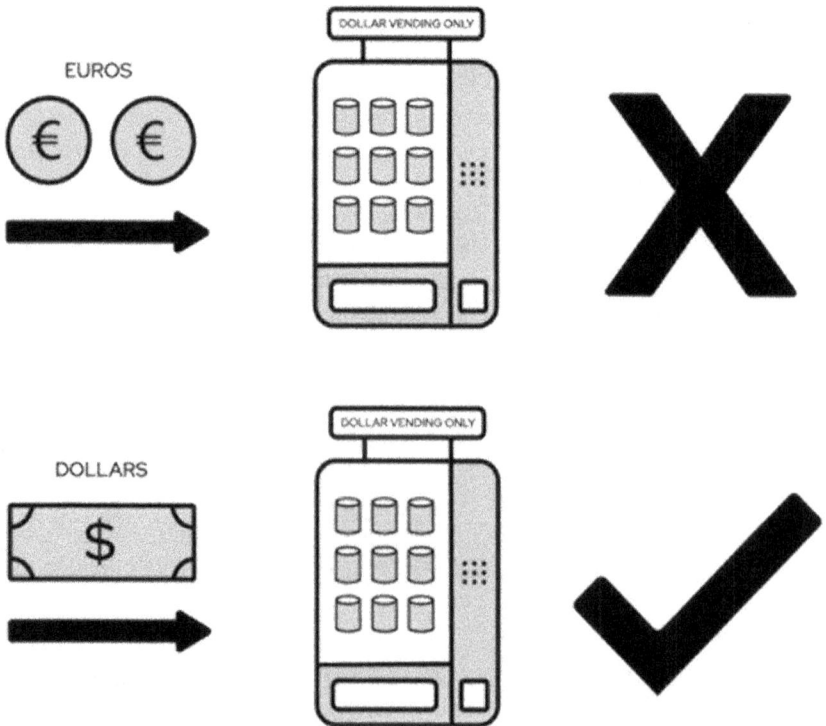

Figure 2.1 Placing the incorrect money into a vending machine will not yield any result, but placing the correct money into the machine will ensure a known outcome, i.e., the drink or snack will be delivered.

key as the **password** to your digital vault – whoever holds the private key has control over the assets inside it.

A **digital wallet** acts as your **identity** and **access point** to the decentralized web. Just as you use a physical wallet to store money and cards, a Web3 wallet holds your digital assets securely and lets you interact with **dApps**, make transactions, and even engage in **DeFi** activities. Popular wallet providers like **MetaMask** and **Phantom** allow users to interact with dApps, sign transactions, and prove their ownership of digital assets – without relying on centralized platforms.

However, unlike traditional financial systems, in Web3 you are responsible for securing your wallet. If you lose access to your private key, you lose control over your assets. This responsibility is one of the key distinctions between traditional financial systems and Web3, where you are your own bank. While this gives you more control, it also means you need to take precautions to secure your keys – keeping them in **secure places** such as **hardware wallets** (physical devices) or **encrypted backups**.

Unknown
"Not your keys, not your crypto."

TYPES OF CRYPTO WALLETS

There are two main types of digital wallets: **custodial** and **non-custodial**.

1. **Custodial wallets**: A **custodial wallet** is one where a third party – such as a cryptocurrency exchange like **Coinbase** or **Binance** – holds the private keys on your behalf. This means that the third party is responsible for the security of your assets.
   While custodial wallets are user-friendly and convenient for beginners, they also introduce **centralization** and the **risk** that the third-party service could be hacked, shut down, or go out of business. In this case, you may lose access to your funds.
2. **Non-custodial wallets**: A **non-custodial wallet**, on the other hand, gives you full control over your private keys. You are the sole owner and custodian of your assets. While this offers greater security and autonomy, it also places the responsibility of securing your assets on you. Non-custodial wallets come with higher levels of **privacy** and **security**, as they are not dependent on a third-party service. Examples of non-custodial wallets include **MetaMask** and **Phantom** (a software wallet) and **Ledger** (a hardware wallet).

When choosing a wallet, you must weigh the trade-offs between convenience and control. If you're new to Web3 and want an easy-to-use option, custodial wallets are a good entry point. However, if you want full ownership of your assets and are willing to take on more responsibility, non-custodial wallets are the better option.

WEB3 USE CASES IN THE REAL WORLD

Web3 is still in its early stages, but it's already beginning to disrupt industries and create new opportunities that weren't possible with traditional centralized systems. From finance to gaming and to art, Web3 technology is bringing about a new era of digital ownership, decentralization, and greater user control. Turning our attention now to real-world examples of how Web3 is being used today and the impact it is having on various sectors we can see some emerging use cases:

1. **DeFi: Reimagining the financial landscape**
    One of the most exciting and impactful applications of Web3 is DeFi. DeFi refers to a suite of financial services – such as lending, borrowing, trading, and yield farming – that operate on blockchain networks, eliminating the need for traditional financial intermediaries such as banks or brokers.

    With DeFi, individuals have access to a range of financial tools that were previously restricted to those within traditional banking systems, providing greater **financial inclusion** and **autonomy**.

    Take **Aave**, a decentralized lending platform, as a prime example. Aave allows users to lend and borrow cryptocurrencies directly, without relying on banks or centralized authorities. For instance, users can deposit their assets, such as Ethereum or stablecoins, into liquidity pools on the platform and earn interest. Conversely, they can take out loans using their crypto holdings as collateral – without the need for a credit check or bank approval. This opens up lending and borrowing to anyone with an internet connection and cryptocurrency, offering financial services to the **unbanked** or those living in regions with limited access to traditional banking.

    What makes DeFi particularly compelling is its **trustless nature:** because blockchain transactions are verified through consensus mechanisms, there's no need for trust in intermediaries. The smart contracts that power DeFi protocols ensure that actions such as loan disbursement or yield payments are carried out automatically, based on pre-established rules and conditions. By cutting out middlemen, DeFi platforms such as Aave provide users with more **competitive interest rates, greater privacy,** and **reduced costs.**

2. **NFTs: Revolutionizing digital ownership and art**

   NFTs, or **non-fungible tokens**, are another powerful example of how Web3 is reshaping industries. NFTs are unique digital assets that represent ownership of a specific item, whether it's artwork, music, video clips, virtual real estate, or even in-game items. Unlike cryptocurrencies like Bitcoin or Ethereum, which are **fungible** (interchangeable with one another), NFTs are **one-of-a-kind** and cannot be replicated, making them ideal for representing exclusive ownership.

   The art world has been **transformed** by NFTs, giving digital artists the ability to sell their work in new ways. **OpenSea**, one of the leading NFT marketplace, allows artists to mint, sell, and trade digital art directly with buyers, without the need for galleries or auction houses. Artists retain **full control** over their creations, ensuring they're compensated directly through smart contracts. Moreover, NFTs have allowed for new forms of **digital scarcity**: limited-edition digital art can be bought, sold, and traded with verifiable ownership attached to the blockchain.

   An added benefit of NFTs is the **royalty feature** – artists can embed smart contracts into their NFTs that ensure they receive a percentage of each subsequent sale. This is a game changer for digital creators, as it ensures they are continually compensated as their work increases in value. Platforms like **Magic Eden** and **OpenSea** also allow creators to set the terms of their sales, providing a direct relationship between artist and buyer.

   Beyond the art world, NFTs are also being used to represent ownership of collectibles, virtual goods, and even as a means of **authentication** for physical products. For example, some luxury brands are exploring the use of NFTs to guarantee the authenticity of their products, allowing consumers to verify ownership and prevent counterfeiting.

3. **Decentralized social networks: Empowering users with control**

   Web3 is also making its mark in the realm of social media, with decentralized social networks offering users more control over their data, content, and interactions. Traditional social media platforms such as **Facebook** and **X** are centralized, meaning they are governed by a single company, which controls how content is shared, how data is used, and how users are engaged. Web3 social networks, on the other hand, are decentralized and run on blockchain technology, meaning no single entity has control.

   Mastodon and **Steemit** are two prime examples of decentralized social networks. Mastodon operates on a federated model, where multiple servers (or 'instances') work together, but each is independently owned and controlled. Users can join different instances based on their interests and communities but retain the ability to interact across the federated network. Steemit, a blockchain-based blogging platform, rewards users for creating and curating content with **STEEM tokens**, allowing users to profit from their posts and interactions.

The main benefit of these decentralized networks is **data ownership**. In traditional Web2 platforms, user data is stored on centralized servers owned by corporations that monetize this data. In decentralized networks, users own their own data and can choose how it is shared, providing a more **private** and **transparent** online experience. Additionally, decentralization eliminates the risk of censorship and content moderation by a central authority, allowing for more diverse and open discussions.

OVERCOMING BARRIERS TO WEB3 ADOPTION

While Web3 presents tremendous potential, there are several barriers that currently prevent its widespread adoption. Overcoming these challenges is essential for ensuring that Web3 becomes a mainstream technology that benefits both individuals and businesses.

1. **Complexity and usability**
    One of the main obstacles to Web3 adoption is **usability**. Unlike Web2 applications, which are designed with user-friendliness in mind, Web3 applications often require a higher level of **technical knowledge**. For example, setting up a cryptocurrency wallet, understanding private keys, and interacting with dApps can be intimidating for beginners.

    To overcome this challenge, developers need to focus on making Web3 interfaces more **intuitive** and **user-friendly**. Platforms like **MetaMask** and **Trust Wallet** have already made strides in simplifying the process of managing digital assets, but there's still a long way to go. Innovations like **Web2.5 wallets**, which combine the best aspects of traditional Web2 and Web3, are helping ease the transition. These wallets allow users to engage with decentralized networks without the need to directly manage private keys, making Web3 more accessible to the masses.

2. **Security and trust concerns**
    Another significant challenge is **security**. In Web3, users are responsible for managing their private keys, which serve as the key to accessing their digital assets. If users lose their private key, they lose access to their assets forever – there's no central authority to recover it. This has led to concerns about **security risks** and **scams**, particularly with the rise of **rug pulls** in the NFT space, where creators vanish with investors' funds.

To try and protect against lost assets and various attack vectors businesses and platforms must prioritize **robust security protocols**, such as **multi-signature wallets** and **two-factor authentication**, to enhance user protection. Education is also critical – informing users about common scams, phishing attacks, and best practices for securing their assets will help build trust in Web3 technologies.

3. **Regulatory uncertainty**

   The regulatory landscape for Web3 technologies, particularly cryptocurrencies and DeFi, is still uncertain. Some countries, such as **El Salvador**, have embraced cryptocurrencies like Bitcoin as legal tender, while others have imposed **restrictions** or outright **bans**. This regulatory uncertainty creates challenges for businesses and individuals looking to engage with Web3 technologies.

   For businesses, navigating this regulatory landscape is critical to ensuring **compliance** with existing laws and avoiding potential legal risks. They must stay informed about regulatory developments and engage legal experts to ensure they're operating within the law. For users, understanding the **legal implications** of Web3 participation, including issues related to **taxation** and **asset ownership**, is crucial for avoiding potential legal issues.

4. **Network scalability**

   As the number of users engaging with Web3 grows, scalability becomes a key issue. Popular blockchain networks like **Ethereum** have faced challenges related to **transaction fees** and **network congestion**. High fees and slow transaction times make Web3 less practical for everyday use, particularly for applications that require fast and affordable interactions, like gaming or e-commerce.

   Other blockchain platforms like **Solana** and **Polygon** offer alternatives with faster transaction speeds and lower fees, making Web3 more viable for mainstream adoption.

THE FUTURE OF WEB3: A VISION FOR TOMORROW

As Web3 continues to evolve, it holds immense potential for the future. The technology promises a more **decentralized** and **user-centric** internet, where individuals have more control over their data, assets, and identities. The future of Web3 is still being shaped, but it's clear that the impact will be profound.

1. **A decentralized web of interoperable applications**

   One of the most exciting prospects of Web3 is the creation of a **truly decentralized** web. In contrast to the current internet, where users are locked into the walled gardens of platforms such as Facebook,

Google, and Amazon, Web3 allows for greater **interoperability** between applications. With blockchain technology, users can take their digital assets, identity, and data across different platforms without losing control.

For example, a user could move their digital identity from one social media platform to another or transfer digital assets such as NFTs from one marketplace to another, all without needing to create new accounts or profiles. This interoperability between platforms could create a seamless, user-centric digital experience where individuals have control over their online presence and interactions.

2. **Web3 and the future of work**

    Web3 is also poised to change the way we work. The rise of **decentralized autonomous organizations (DAOs)** is a key example of how Web3 is redefining the workplace. DAOs are organizations that are run by smart contracts and governed by token holders rather than a central authority. These organizations allow for more demo- cratic decision-making, where participants have a direct say in the future direction of the project – we are going to look into DAOs in a subsequent chapter with far more depth.

    In the future, we may see **DAOs** becoming more common, with workers contributing to projects on a decentralized basis. This could lead to the rise of a **global workforce** where individuals can collab- orate on projects from anywhere in the world, without being tied to a specific company or geographical location.

3. **The integration of Web3 with other emerging technologies**

    Web3 is not an isolated phenomenon. It is part of a broader wave of innovation that includes technologies like **artificial intelligence (AI), machine learning (ML), advanced cloud computing,** and **the Internet of Things (IoT)**. In the future, we can expect to see **Web3 applications** integrated with these technologies to create **smarter, more efficient digital ecosystems**.

    For example, AI-powered dApps could provide users with personalized recommendations based on their preferences and behavior. Smart cities could leverage blockchain and IoT to create decentralized, autonomous infrastructures that improve the quality of life for residents while reducing reliance on central authorities. We are also in the early stages of the adoption of AI agents built on Web3 technology and expect this to increase, and our later chapter will share views on this.

4. **Transforming industries: From finance to healthcare**

    Finally, Web3 has the potential to overhaul entire industries, creating more transparent, secure, and efficient systems. In **finance**, we've already seen the rise of DeFi platforms, which provide an alternative to traditional banking. But Web3's impact won't stop there. We could see Web3 disrupting industries like **healthcare,** where blockchain can be used to securely store and share patient data, ensuring privacy and reducing administrative costs.

Similarly, in the **supply chain** industry, blockchain can be used to provide greater transparency and traceability, reducing fraud and ensuring that products are ethically sourced and transported. In **intellectual property** law, Web3 can help artists and creators protect their work and ensure that they are compensated fairly. The limits and possibilities are endless.

And so, whether it's through the creation of **interoperable dApps,** the **integration with emerging technologies** like AI and IoT, or the **disruption of industries** like finance and healthcare, Web3 is setting the stage for a new digital era.

As adoption grows and challenges are overcome, we'll likely see Web3 become a cornerstone of the internet's next generation and transform industries along the way. Are you ready?

ACTIONABLE STEPS : HOW TO GET STARTED WITH WEB3

For those looking to enter the world of Web3, the first step is to familiarize yourself with the basic tools and technologies. Here are some **practical steps** to get started:

1. **Create a crypto wallet**: To interact with Web3, you'll need a digital wallet. Choose a wallet that fits your needs – whether it's a non-custodial wallet like **Phantom** or a custodial wallet like **Coinbase**. Remember to store your private keys securely and never share them with anyone.
2. **Engage with dApps**: Start exploring dApps. There are plenty of dApps for finance (e.g., **Uniswap, Aave, Jupiter, MarginFi**), NFTs (e.g., **OpenSea, Magic Eden, Tensor**), and gaming (e.g., **Axie Infinity**). These platforms allow you to experience Web3 firsthand and begin using decentralized technologies.
3. **Educate yourself**: The Web3 space is rapidly evolving, so it's important to stay informed. Follow Web3 influencers, read whitepapers, and participate in online communities to keep up with the latest developments.
4. **Experiment with NFTs**: If you're interested in art or digital ownership, consider buying and selling NFTs. Start by using platforms like

OpenSea to mint or purchase NFTs, and experiment with owning and trading digital art.

5. **Join a DAO**: For those interested in decentralized governance and collaborative decision-making, joining a **DAO** can provide valuable experience in how decentralized organizations function.

Chapter 3

Blockchain unleashed
The backbone of Web3

As technology changes, upgrades, and new solutions arrive, it is blockchain technology that stands out as one of the most revolutionary innovations of the 21st century. While the term 'blockchain' is often associated with cryptocurrencies like Bitcoin, its potential extends far beyond digital currencies. Whilst we will focus in this chapter a lot on those cryptocurrency use cases, we will share with you the full scale of the impacts blockchain will have.

Blockchain technology is reshaping the way we interact with data, manage transactions, and build trust in digital systems. Understanding the blockchain is crucial for anyone looking to navigate the Web3 ecosystem, as it serves as the foundational technology enabling decentralization, transparency, and trust in a variety of applications – from finance to supply chains to healthcare as we saw at the end of the previous chapter.

Remember that the blockchain is the engine that powers the next generation of the internet, Web3, where users gain greater control over their data, identities, and digital assets. Unlike Web2 platforms, where centralized entities control users' data, blockchain enables a **peer-to-peer** (**P2P**) network, allowing for decentralized applications (dApps) that empower users. Its potential to replace outdated, centralized infrastructure in nearly every industry is why blockchain is often described as a **paradigm shift** in how we think about digital transactions and interactions. This is what we are going to cover now.

> *"The blockchain is the biggest innovation since the internet."*
> Marc Andreessen (co-founder of Andreessen Horowitz)

DOI: 10.1201/9781003616504-3

BLOCKCHAIN BASICS: UNDERSTANDING THE CORE TECHNOLOGY

Blockchain technology was first made popular in 2009 with the creation of **Bitcoin,** a decentralized digital currency designed to solve the problem of **double-spending** in digital transactions. **Double-spending** is the issue where a digital asset (like a file or token) could be duplicated and spent more than once, which was not a problem with physical currency, but posed as a challenge for digital money – with Bitcoin this was simply made impossible in code, and you can see this illustrated in Figure 3.1. This challenge was brought even more to the forefront of people's minds in the wake of the 2008 financial crisis where the mismanagement of digital money became painfully apparent to many.

However, Bitcoin's creation was just the starting point. The underlying technology of blockchain that enabled Bitcoin is far more powerful and versatile than just a currency. Blockchain technology, since its inception, has expanded to become the foundation for **Web3** and the building blocks of a new generation of digital applications. Today, blockchain has evolved into a multi-faceted technology supporting not just cryptocurrencies, but **smart contracts, decentralized finance (DeFi), supply chain transparency,** and much more.

Blockchain's versatility and power come from its ability to provide a **secure, immutable** way to store data, making it ideal for various applications across multiple industries. Let's break down the basics of blockchain technology and explore why it has become the **backbone of Web3.**

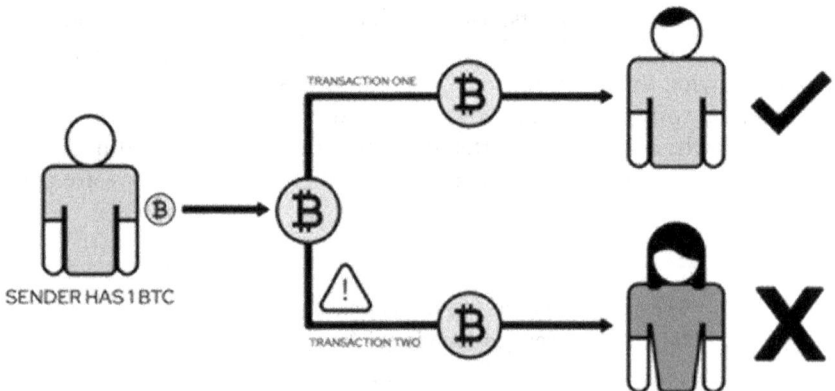

Figure 3.1 Bitcoin cannot be duplicated, and thus the same Bitcoin cannot be sent at the same time in two separate transactions to two separate receivers (i.e., duplicated). The traceability and security of the blockchain means it is not possible to double-spend.

THE ARRIVAL OF BITCOIN:
A GROUNDBREAKING MOMENT

In 2008, **Satoshi Nakamoto**, the mysterious figure (or group of figures) behind Bitcoin, introduced blockchain to the world through a groundbreaking whitepaper titled '**Bitcoin: A Peer-to-Peer Electronic Cash System**.'

In this paper, Nakamoto outlined how Bitcoin could be used to solve the problem of **double-spending** in digital finance. Double-spending occurs when digital assets can be copied and used multiple times, which creates a major flaw in digital currencies. Traditional financial systems solve this problem by using a **centralized authority**, like a bank, to validate transactions. However, Nakamoto's innovation was to use a **decentralized network of computers** to validate transactions – removing the need for any central authority to ensure trust.

Bitcoin's use of blockchain technology allowed for a **P2P** network of participants to verify and validate transactions without relying on any centralized party. This enabled a **trustless** system, where users could transact directly with each other, without needing to trust a third party like a bank. Instead of relying on a central entity, Bitcoin leveraged the collective effort of **miners** and **validators** within the network to verify transactions and secure the blockchain.

Bitcoin's success demonstrated the potential of blockchain technology to revolutionize the way we think about money, trust, and digital interactions. But Bitcoin was just the first application. As we'll explore, we will find blockchain's potential stretches far beyond digital currencies and into a range of industries that are beginning to embrace its **decentralized, transparent**, and **secure** features.

"At the end of the day, Bitcoin is programmable money. When you have programmable money, the possibilities are truly endless."
Andreas M. Antonopoulos (author)

WHAT IS BLOCKCHAIN?

At its core, a **blockchain** is a **decentralized digital ledger** that records transactions across multiple computers in such a way that the record cannot be altered retroactively without the alteration of all subsequent blocks. This decentralized nature ensures that **no single entity** or individual has control over the entire database. This makes blockchain not only **secure** but also **transparent** and **trustworthy**, which is a fundamental principle behind **Web3** – a decentralized web where users control their data, interactions, and digital identities.

Each **block** in the blockchain contains a list of **transactions**. These transactions could include anything from a cryptocurrency exchange to the transfer of digital assets, voting information, or any form of data that needs to be securely recorded. The key is that **each block is cryptographically linked** to the previous block, forming a continuous **chain** – hence the term **blockchain** (as shown in Figure 3.2). Once a block is added to the chain, it is nearly **impossible to change** the information within it without invalidating the entire structure. This makes blockchain an incredibly **trustworthy** and **tamperproof** system, ideal for recording valuable information in an immutable and transparent way.

The decentralized nature of blockchain means that transactions are validated and recorded by multiple computers, called **nodes**, rather than a single central authority. In traditional centralized systems, like banks, all transactions are processed and validated by a single entity. This system relies on trust; users trust the bank to ensure their funds are secure, and the bank trusts its own systems.

However, in blockchain, transactions are validated by a **distributed network of computers**, making the entire system transparent and resistant to manipulation. This decentralization removes the need for a **trusted third party** to validate transactions, creating a more **efficient, secure**, and **trustless** environment for conducting business.

Each block contains **cryptographic hashes**, which are essentially **digital fingerprints** of the data inside the block. These hashes link each block to the next in the chain, ensuring that once a block is added to the blockchain, it is part of a permanent record. If an attacker were to attempt to modify a block, they would have to change the **cryptographic hash** of that block, and in turn, every subsequent block would also need to be altered. This makes blockchain extremely resistant to fraud or tampering.

Figure 3.2 Blocks linked to one another in order to form a chain: known as a blockchain.

HOW DOES BLOCKCHAIN WORK? (IN SIMPLE TERMS)

Let's simplify it a little before we get into the details and technical elements.

Imagine you and your friends are playing a game where you keep track of your scores in a notebook. Each time someone scores, they write it down in the notebook. But, there's one rule: no one can erase or change anything once it's written. If someone tries to change their score, everyone will see it and oppose the change.

Now, let's make this game more interesting and digital. Instead of a notebook, we'll use something called a **blockchain**. Blockchain is like a digital notebook, but much safer, more transparent, and much harder to tamper with.

Here's how it works, step by step:

1. The digital ledger (the notebook)

In the blockchain, instead of one person writing in a notebook, everyone has a copy of the same **digital notebook**. This notebook doesn't belong to any one person, and everyone who participates in the game gets to see and update it. This is important because it means no one can secretly change the score without everyone noticing.

2. Adding transactions (writing down the scores)

Let's say you score a point in the game. You tell everyone your score, and everyone writes it down in their copy of the notebook at the same time. This score is called a **transaction**.

A **block** is like a page in this notebook where a group of transactions (scores) are written down together. Each time a new group of scores is added, a **new block** is created.

3. The block is linked together (the chain)

Now that your score is added to the notebook's page (block), it's time to link it to the previous page of the notebook (previous block). Each page (block) has a **special number** (called a cryptographic hash) that is like a digital fingerprint of the previous page.

This fingerprint ensures that once a page (block) is added, it's **linked to the one before it** in a very strong way. If anyone tries to change a score on an earlier page, the fingerprint will no longer match, and the new page won't connect properly. This makes the entire chain unbreakable.

So, the pages are linked together to form a **chain of blocks** – that's why it's called a **blockchain**.

4. Why is it so secure? (immutability)

Let's say someone tries to sneakily change their score on one of the pages after it's been written. To do this, they would not only have to change their page (block), but they would also need to change every page that came after it! Why? Because each page has that special fingerprint of the previous page. Changing one page would mess up the entire chain, and everyone would notice.

That's why once a page is added, it's very hard to change it. This makes the blockchain **secure** and **tamperproof**. It's like having a notebook written in permanent ink that simply cannot be edited.

5. Decentralization: No one controls the blockchain

Here's where it gets even cooler. Unlike a normal notebook where one person could be in charge of writing, in the blockchain, **everyone** who participates has a copy of the notebook. This means no single person can control or change the data by themselves.

Every time a new page (block) is added, the whole network checks to make sure it's valid. This process is done by many computers (called **nodes**) in the network. They work together to confirm that everything is correct before adding the page to the notebook.

6. Example: Sending money with blockchain (crypto)

Let's say you want to send $10 to your friend using a blockchain system like **Bitcoin**. When you send the money, it creates a transaction. This transaction is added to the blockchain as a block, and everyone in the network verifies that it's real (you have $10 and haven't spent it elsewhere). Once confirmed, the transaction is added to the blockchain, and your friend now has $10 in their wallet.

Because the blockchain is decentralized and everyone has a copy of it, there's no need for a bank or middleman to verify the transaction. It's all done by the network, and the information is visible to everyone involved. Plus, once it's added, it can't be changed or erased – your friend will always have that $10 in their wallet unless they spend it.

7. Summary: Why blockchain is important

- **Secure**: Once information is added, it's nearly impossible to change.
- **Transparent**: Everyone involved can see the data and verify transactions.

- **Decentralized**: No one person or company controls it. Everyone shares a copy.
- **Tamperproof**: If someone tries to mess with the data, it would be noticed immediately by everyone.

Blockchain isn't just about cryptocurrencies like Bitcoin – it can be used to make an autonomous financial system, a censorship free network and so much more.

So, now that we understand what Blockchain is and how it works, let us turn our attention to what makes it really special and unique in more depth.

1. **Blockchains Innovation: Immutability and Transparency**
   Blockchains have many unique features that make them innovative and useful. The key innovation that makes blockchain so powerful is its **immutability** – once data is recorded on the blockchain, it cannot be altered or deleted. This is achieved through **cryptographic hashing** and **consensus mechanisms**.

   When a block is created, it receives a **cryptographic hash**, which is a unique string of characters that represents the data in the block. This hash is then included in the subsequent block, forming a continuous chain of blocks. The cryptographic hashes link all blocks together, and if an attacker were to try and alter the data in one block, the hash would no longer match, thus disrupting the entire chain.

   This **immutability** ensures that once data is written to the blockchain, it is permanent, verifiable, and transparent. It provides a level of **security** and **trust** that traditional centralized databases cannot offer. Because the blockchain is decentralized and transparent, **all participants in the network can see the same data** – there is no need to trust a central authority, as the blockchain's structure ensures the integrity of the data through the collective validation of the network.

   In addition, blockchains transparency makes it an ideal solution for applications where accountability is crucial – whether that's tracking the **origin** of products in a supply chain, ensuring the **authenticity** of digital assets like NFTs, or verifying the **accuracy** of voting systems. The blockchain provides a transparent, tamperproof record of all transactions, allowing for greater **trust** and **accountability** in industries and systems that depend on verified information.

2. **The Decentralized Nature of Blockchain: No Central Authority**
    In traditional systems, central authorities (like banks, government entities, or corporations) control the flow of information and transactions. In contrast, blockchain operates on a **P2P network**, where all participants have an equal say in validating and recording transactions. This decentralization eliminates the need for intermediaries, reducing costs and increasing efficiency.

    The beauty of decentralization is that it allows for **trustless transactions** – participants don't need to trust a central authority because the blockchain ensures that the data is accurate and immutable. This is particularly powerful in industries like **banking**, where trust and security are paramount. Blockchain technology enables secure transactions without the need for a third party to verify the integrity of the transaction.

> *"The blockchain does one thing: It replaces third-party trust with mathematical proof that something happened."*
> Adam Draper (investor)

---

3. **Cryptography: The Foundation of Blockchain Security**
    The security of blockchain relies heavily on **cryptographic techniques**. At the core of blockchain security is **cryptographic hashing**, a process that converts input data (like a transaction) into a fixed-length string of characters. Each block contains a hash of the previous block, which links the blocks together and ensures that the chain cannot be altered once it's been created.

    Another key aspect of blockchain security is **public and private key cryptography**, which allows users to sign transactions and prove ownership of digital assets. Each user has a **public key** (similar to an email address) and a **private key** (like a password). The public key is used to receive assets, while the private key is used to authorize transactions and prove ownership.

    These cryptographic principles ensure that blockchain transactions are secure, verifiable, and resistant to tampering, as you can see in Figure 3.3 that shows the process of opening a document (data) only by having the relevant keys in place.

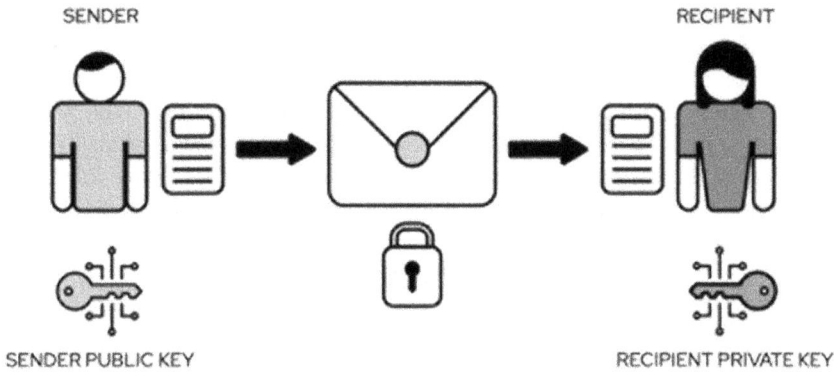

SENDER   RECIPIENT

SENDER PUBLIC KEY   RECIPIENT PRIVATE KEY

Figure 3.3 The use of a public key and a private key to send and receive a document on the blockchain, without the ability to tamper or edit that document.

REAL-WORLD BLOCKCHAIN USE CASES: TRANSFORMING INDUSTRIES AND SYSTEMS

Blockchain's potential extends far beyond cryptocurrencies like Bitcoin. As we move into the era of Web3, blockchain technology is beginning to reshape industries such as healthcare, supply chain management, voting systems, digital identity, and more. These use cases highlight blockchain's versatility, security, and ability to disrupt traditional systems.

1. Blockchain in healthcare: Ensuring data security and interoperability

Blockchain's potential to revolutionize healthcare lies in its ability to securely store, track, and share sensitive medical data. Traditional healthcare systems often rely on siloed databases maintained by various hospitals, clinics, and healthcare providers, creating inefficiencies and potential risks when it comes to patient care. Patient data, like medical histories, treatment plans, and prescription details, is often fragmented across multiple platforms, leading to incomplete or outdated records. This fragmentation can delay care, create room for errors, and prevent smooth collaboration among healthcare professionals.

Blockchain can solve these problems by creating a **single, immutable source of truth** for all medical data. This is achieved through a **decentralized ledger** where patient data is securely stored on the blockchain, with every update made by healthcare providers recorded in a way that cannot be tampered with.

For example, **Solve.Care**, a healthcare platform powered by blockchain, ensures **data security**, **privacy**, and **interoperability** by allowing doctors, patients, and insurance companies to securely share and access healthcare information. With blockchain, patients can have full control over their data, determining who has access and for how long, and giving them the ability to revoke access if necessary. This improves the efficiency of healthcare delivery and ensures that important information is accessible when needed.

A practical example is a patient visiting a specialist in a different city or country. With blockchain, their health records could be securely shared with the new doctor, who would then have access to the patient's complete medical history, reducing the risk of misdiagnosis and improving the quality of care.

2. Blockchain in voting systems: A secure, transparent solution

Voting systems in many countries remain vulnerable to **fraud, tampering**, and **manipulation**, which undermines the integrity of elections. While digital voting systems have been introduced to simplify and increase participation in elections, they are often plagued by **security vulnerabilities**. These vulnerabilities include issues like hacking, vote tampering, and the possibility of election fraud. Blockchain technology has the power to solve these problems by making voting **transparent**, **secure**, and **tamperproof**.

Take **Estonia's digital voting system** as an example – which is in trial, and certainly not the final solution. Estonia has been a leader in integrating digital solutions into its governmental services, including blockchain-based e-voting. With Estonia's **blockchain-powered system**, every vote is cryptographically recorded on the blockchain, ensuring that it cannot be altered once cast. Moreover, the **voter's identity** is protected, maintaining anonymity while ensuring transparency in the election process.

The blockchain's immutable ledger records every action, from the casting of the vote to the counting and reporting, making the entire process open and verifiable by all participants. This significantly reduces the chances of **fraud** and improves **public trust** in the democratic process.

Using blockchain in voting could have massive global implications, enabling countries to adopt **digital voting** systems that are not only more efficient but also **safer** and more **transparent**. It could lead to more accessible voting, lower costs for elections, and a greater overall turnout.

3. Blockchain in digital identity: Empowering users with control

In our digital age, most people's personal information – such as their **name**, **address, date of birth**, and **social security number** – is controlled by centralized companies like **Google**, **Facebook**, and **banks**. While this has allowed for some convenience, it also presents serious risks related to **privacy, security**, and **identity theft**. Data breaches have exposed the personal information of millions of people, leading to identity theft, fraud, and the misuse of sensitive information. Blockchain technology offers a solution by **empowering users** to **control their own identities**.

The **Sovrin Foundation** is developing a **self-sovereign identity system** based on blockchain. In this system, individuals create their digital identity on the blockchain, which is securely stored and cryptographically verified. Rather than relying on third-party entities to confirm identity (such as showing a physical ID to prove your age), you can use **blockchain-based credentials** to verify who you are.

The key benefit of this model is that individuals have full control over their personal data. For example, if someone needs to prove their **age**, they can share just the necessary information from their blockchain identity, without revealing any extra personal details. This reduces the risk of **data breaches** and ensures that individuals have **greater privacy** and **security**.

In a world where centralized platforms often exploit and sell user data, blockchain-based **self-sovereign identities** could drastically change how we think about **privacy** and **control** in the digital world.

4. Blockchain in supply chain management: Transparency and efficiency

Supply chains are often complex and fragmented, involving multiple parties, intermediaries, and processes to get a product from the manufacturer to the consumer. This complexity can lead to inefficiencies, fraud, and an inability to track the provenance of goods. For instance, it's difficult for consumers or businesses to know exactly where a product came from, whether it was ethically sourced, or whether it meets quality standards.

Blockchain can address these issues by providing a transparent, **tamperproof** record of every transaction in the supply chain. For instance, **Walmart** has partnered with **IBM** to use blockchain to improve its food safety efforts. Using blockchain, Walmart can trace the origin of food products from farm to table, providing complete transparency on where the food came from, how it was handled, and what processes it went through before reaching consumers.

For example, if there is a **foodborne illness outbreak**, Walmart can trace the source of the contaminated product in seconds, instead of days or weeks, drastically improving the efficiency of product recalls. This **transparency** not only improves safety but also reduces fraud, ensures **ethical sourcing**, and enhances **consumer trust** in the brand.

Companies like **Nestlé** and **Unilever** are also experimenting with blockchain to ensure product traceability in supply chains. As blockchain allows companies to track every step a product goes through, it could lead to **more responsible consumption**, greater **corporate accountability**, and **better-informed consumers**.

5. Blockchain in intellectual property: Protecting creators' rights

In the digital world, creators often struggle with protecting their **intellectual property (IP)** – be it **art**, **music**, or other forms of creative content. The rise of digital platforms has made it easier to distribute content globally, but it has also made it easier for content to be **pirated** or used without permission. Blockchain technology offers a solution by providing **clear ownership records** and **authenticity tracking** that can help creators protect their work.

This technology could radically change the creative industries by making it easier for creators to maintain **control over their IP** and make sure they are properly compensated for their work. Additionally, it provides a **transparent** and **secure** method to track the provenance of digital art, which could help fight **piracy** and **unauthorized distribution** of content.

6. Blockchain in the energy sector: Revolutionizing energy trading

The traditional energy sector is often inefficient, with centralized systems that rely on a few large companies to generate, distribute, and sell energy. This creates challenges, including high costs, lack of transparency, and inefficiencies in energy trading. Blockchain offers a new way of trading energy by using **P2P networks** to allow individuals and businesses to directly buy and sell excess energy.

Power Ledger, an Australian company, is using blockchain to create a decentralized energy marketplace where individuals can **buy and sell solar energy** directly with each other. Through the platform, if someone produces more energy than they need (for instance, from a rooftop solar panel), they can sell it directly to someone else who needs it, bypassing traditional utility companies.

This blockchain-based **energy marketplace** helps increase **energy efficiency** by allowing for the **democratization of energy** – putting power back into the hands of consumers. It also helps drive **sustainability** by encouraging the use of renewable energy sources. Consumers now have more control over their energy consumption, and they can **trade energy** in a **transparent**, **secure**, and **efficient** way.

THE ROLE OF CONSENSUS MECHANISMS IN BLOCKCHAIN

For a decentralized network to work effectively, it needs a mechanism to ensure that everyone agrees on the contents of the ledger. In traditional systems, a central authority (like a bank) decides what's valid or invalid, ensuring all records are accurate. But in blockchain, consensus must be achieved by the network of participants, who are typically called **nodes** or **miners** (in the case of proof-of-work systems like Bitcoin).

This is where **consensus mechanisms** come in. These are protocols that ensure all participants agree on the current state of the blockchain. Essentially, a consensus mechanism helps the network come to an agreement on which transactions are valid and true on the blockchain, ensuring the integrity of the data.

There are several types of consensus mechanisms, but the two most common are **Proof of Work (PoW)** and **Proof of Stake (PoS)**.

- **PoW**: In PoW, participants, known as **miners,** compete to solve complex mathematical puzzles. The first one to solve the puzzle gets the right to add a new block to the blockchain. This process requires a significant amount of computational power and energy, which is why it's often criticized for its environmental impact. However, it remains one of the most secure consensus mechanisms and is used by Bitcoin and other cryptocurrencies.
- **PoS**: PoS is a more energy-efficient consensus mechanism. In PoS, validators (or 'stakers') lock up a certain amount of cryptocurrency as collateral. The network randomly selects one of the validators to add the next block to the blockchain. The more cryptocurrency a validator stakes, the higher their chances of being selected. PoS is seen as a more sustainable alternative to PoW, and it's used by networks like **Cardano**.

The goal of these mechanisms is to prevent any single entity from taking control of the blockchain. They ensure that no one can alter the ledger without the majority of the network agreeing to the changes. It's important to note that consensus mechanisms do not rely on the goodwill of network participants to maintain integrity, the system is designed to appeal toward economic rationality. Positive actions within the network are always incentivized economically ensuring the goals of both the individual participants and the entire network are aligned.

THE BYZANTINE FAULT TOLERANCE: ENSURING CONSENSUS IN A DECENTRALIZED WORLD

To truly appreciate how blockchain operates, it's important to understand the concept of **Byzantine Fault Tolerance (BFT)** – as shown in Figure 3.4. This term comes from the **Byzantine Generals Problem,** a thought experiment that describes the challenge of achieving consensus in a decentralized system with faulty or malicious actors.

The problem is this: imagine a group of Byzantine generals who must coordinate an attack on a city. The generals communicate by sending messengers, but some of the messengers may be traitors trying to sabotage the plan. The challenge is for the generals to agree on a strategy despite the possibility of some of them being unreliable or dishonest.

In the context of blockchain, **BFT** refers to the system's ability to function and reach consensus even if a portion of the participants are dishonest or fail to communicate properly. As long as the majority of the participants are honest, the system will still be able to reach an agreement and ensure the integrity of the blockchain.

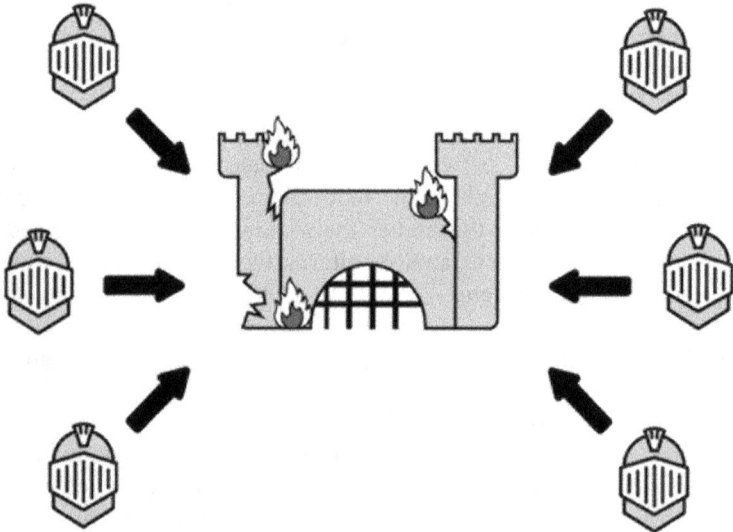

Figure 3.4 The Byzantine Fault Tolerance: A set of knights attack the castle (representing generals), however they fail to agree on the strategy and approach and instead have to retreat, leaving the castle intact and those knights perishing.

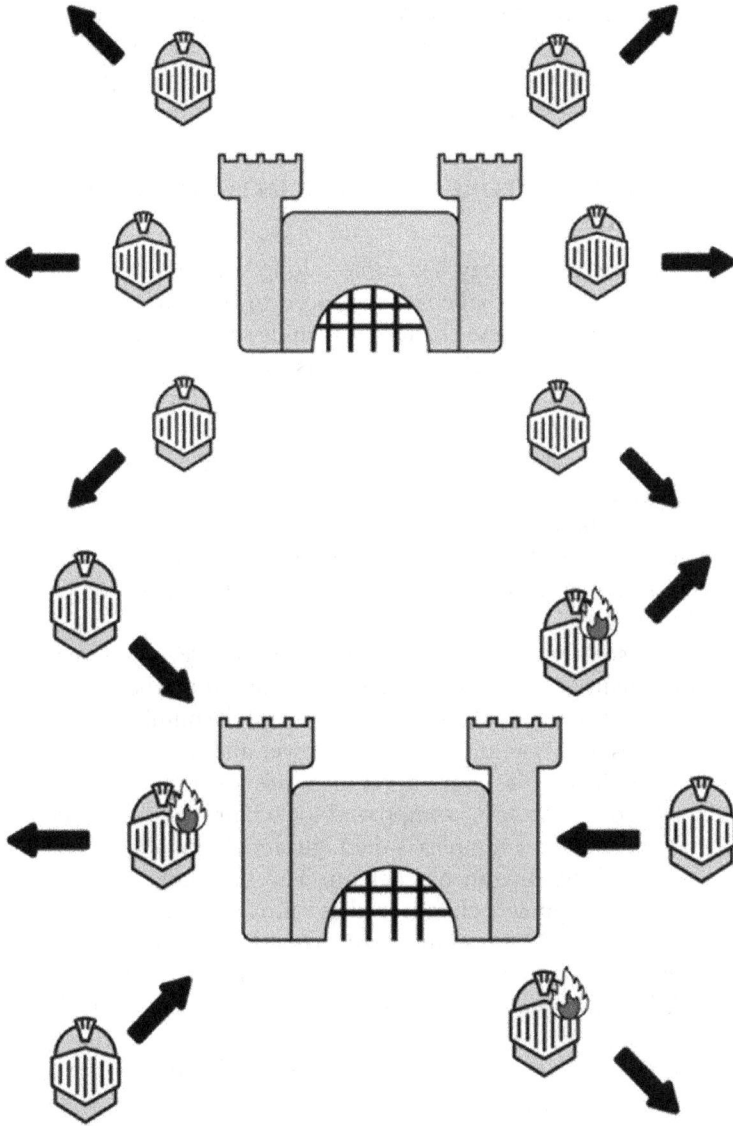

Figure 3.4 (Continued)

This concept is critical in decentralized systems, where no single party is trusted to manage the data. It ensures that the blockchain remains secure and functional, even if some participants are malicious or make mistakes. It's one of the reasons why blockchain is so powerful in providing trust without relying on a central authority.

BLOCKCHAIN'S FUTURE: SCALING, INTEROPERABILITY, AND BEYOND

While blockchain technology has come a long way since the introduction of Bitcoin, there are still challenges that need to be addressed before it can reach its full potential. These challenges include scalability, interoperability, and the adoption of blockchain in various industries. Let's explore these issues and the potential solutions that are being developed:

- **Scalability: Making blockchain fit for the masses.** One of the biggest criticisms of blockchain technology is its **scalability** – the ability to handle a large number of transactions per second (TPS). Bitcoin, for example, can only handle around **7 TPS**, which is a far cry from centralized payment systems like Visa, which can handle **65,000 TPS**.

  The scalability problem arises because every transaction on a blockchain needs to be verified and recorded by the entire network. As more participants join the network and the number of transactions increases, the system can become slower and more expensive.

  To address this, some developers are working on several **Layer 2 solutions** that aim to improve blockchain scalability. One of the most prominent examples is the **Lightning Network**, which operates on top of the Bitcoin blockchain. The Lightning Network creates off-chain payment channels that allow for faster and cheaper transactions. These transactions are later settled on the main blockchain, reducing congestion and improving scalability.

  Another promising solution is the shift to **PoS**, which is not only more energy-efficient but also offers the potential for higher throughput compared to PoW. Ethereum improved the network's scalability and reduced its environmental impact by shifting from PoW to PoS in 2022.

- **Interoperability: Connecting blockchains across networks.** As blockchain technology continues to evolve, **interoperability** between different blockchain networks is becoming an increasingly important issue. Currently, most blockchains operate in isolation, meaning that assets and data on one blockchain cannot easily be transferred to another. This creates a fragmented ecosystem, where the full potential of Web3 cannot be realized.

The solution to this problem lies in the development of **cross-chain technologies**, which enable different blockchains to communicate with each other. For example, the **Polkadot** network aims to create an interconnected web of blockchains, allowing them to share information and assets. Similarly, **Cosmos** is building a decentralized network of blockchains that can interoperate with one another.

By enabling interoperability, Web3 applications can leverage the strengths of multiple blockchains, creating a more flexible and resilient ecosystem. This will be crucial for the mass adoption of dApps and for achieving the vision of a truly decentralized internet.

> *"Blockchain is the tech. Bitcoin is merely the first mainstream manifestation of its potential."*
> Vitalik Buterin (co-founder of Ethereum)

THE FUTURE OF BLOCKCHAIN AND WEB3

The future of blockchain is bright, but there are still many challenges to overcome. As blockchain technology matures, we will likely see it integrated into a wider range of industries, including **healthcare, finance, supply chain management,** and **digital identity.** Blockchain has the potential to transform how we store and manage data, offering a more secure, transparent, and user-centric alternative to the centralized systems that dominate today's internet.

Additionally, the rise of **smart contracts** and **DeFi** – as we will see in the next chapter – is creating new opportunities for individuals and businesses to participate in the digital economy without relying on traditional banks or financial institutions. **dApps** are already disrupting industries such as gaming, social media, and content creation, and they will continue to play a key role in the development of Web3.

While the road ahead is full of obstacles, the potential rewards are immense. By embracing the decentralized future, businesses and individuals can unlock new opportunities, create more equitable systems, and participate in the next generation of the internet.

ACTIONABLE STEPS : GETTING STARTED WITH BLOCKCHAIN

1. **Start learning the fundamentals of blockchain**

   - To begin your blockchain journey, understanding the core principles is crucial. Consider taking online courses or reading introductory books on blockchain and decentralized technologies.
   - Platforms like **Coursera, Udemy**, and **edX** offer excellent beginner courses that break down complex concepts into digestible lessons. This foundational knowledge will equip you with the tools to dive deeper into blockchain applications and use cases.

2. **Experiment with blockchain-based applications**

   - We suggest the best way to learn is by using your digital wallet (which you setup in the previous chapter). Start exploring the world of decentralized applications (dApps). Examples include **Uniswap, Jupiter** (a decentralized exchange), **OpenSea, Magic Eden** (an NFT marketplace), and **Audius, DripHaus** (a decentralized music platform).
     - These platforms will give you hands-on experience in using blockchain for a variety of use cases, whether it's trading cryptocurrency, buying digital art, or interacting with decentralized content.

3. **Join blockchain communities**

   - Blockchain is a rapidly evolving field, and the best way to stay informed is by engaging with the community. There are numerous online forums, social media groups, and networks where blockchain enthusiasts, developers, and innovators share ideas, discuss trends, and provide support. Platforms like **Reddit, Discord**, and X (formerly Twitter) are great places to start. Join these communities to gain insights, ask questions, and collaborate with like-minded individuals (though take precautions to avoid scams, which can be frequent).

4. **Consider blockchain for business innovation**

   - If you are a business owner or professional, think about how blockchain could be integrated into your business. Could you implement a solution to improve transparency or reduce costs? Whether it's improving supply chain tracking, enhancing digital identity verification, or creating DeFi solutions, the possibilities are vast.

Start exploring how blockchain can offer innovative solutions within your industry.

5. Stay informed about emerging blockchain developments

- Blockchain is constantly evolving, and new solutions, platforms, and regulations are emerging every day. To stay ahead of the curve, follow thought leaders in the blockchain space, read whitepapers, and attend webinars or conferences. Blockchain is an exciting field to be a part of, and staying informed will help you make the most of its opportunities.

The future of blockchain: A decentralized world awaits

The shift toward a decentralized future is already in motion, with blockchain technology playing a central role. As adoption grows, its potential continues to expand. Whether you're just beginning to explore blockchain or considering how it fits into your personal or business strategy, now is a good time to get involved.

The coming years will be key in shaping the Web3 landscape. Engaging with blockchain today can help you stay ahead of these developments. Decentralization offers new possibilities – taking the first steps now can help you make the most of them.

Chapter 4

Crypto and DeFi

Redefining the future of finance

Cryptocurrency and decentralized finance (DeFi) are not just trends; they are a revolutionary force within the modern financial ecosystem, challenging centuries-old systems and offering new opportunities to those previously excluded from traditional financial services. At its core, cryptocurrency allows for peer-to-peer transactions of digital value across borders without relying on a central intermediary, such as a bank or a payment processor. This is an innovation that makes transferring money faster, cheaper, and more secure.

But DeFi takes this a step further. It's not just about transferring value – DeFi is about reconstructing entire financial systems. Imagine an ecosystem where financial services like lending, borrowing, saving, trading, and insurance happen without traditional banks or central authorities. Instead, these services are governed by code and executed through smart contracts, removing intermediaries and allowing anyone, anywhere, to participate in financial markets. Through decentralized networks, anyone with an internet connection can access, manage, and grow their assets, without the need for approval from a centralized institution.

The impact of these innovations is profound. They democratize finance, offering a way for individuals in developing nations or marginalized communities to access banking, saving, and investment services that were once out of reach. Furthermore, DeFi introduces a level of **transparency, security, and efficiency** that the legacy financial system simply cannot match. By eliminating the middlemen, DeFi reduces transaction fees, accelerates processing times, and removes the need for trust in any central authority. Every transaction is recorded on a blockchain – a decentralized and immutable ledger – that provides both security and transparency, ensuring trust between users and systems.

In this chapter, we will look into and explore the world of cryptocurrency and DeFi: examining how they challenge traditional finance (TradFi), where they succeed, and the issues they face.

DOI: 10.1201/9781003616504-4

From understanding the basics of digital currencies like Bitcoin to exploring the full potential of DeFi systems, we will break down how these technologies work, the fundamental principles that drive them, and their far-reaching implications for the future of global finance. You'll gain insights into the specific risks and rewards associated with crypto and DeFi and how businesses and individuals can leverage these technologies to unlock new opportunities.

The future of finance is rapidly being redefined, and with that, comes a new era of financial inclusion, transparency, and innovation. By understanding these disruptive forces, you'll be prepared to navigate this evolving landscape and potentially transform how you manage, invest, and interact with money in a decentralized world.

WHAT IS CRYPTO AND DEFI?

Cryptocurrency and DeFi are two key pillars of the Web3 movement, and they are deeply interconnected. To understand their significance, it's crucial to first grasp their basic components and how they work together.

Cryptocurrency is a form of digital or virtual currency that utilizes cryptography to secure transactions, control the creation of new units, and verify the transfer of assets. Unlike traditional currencies, which are controlled by central banks or governments, cryptocurrencies operate on decentralized networks, which means they are not issued or controlled by any central authority.

Bitcoin, the first cryptocurrency, was created in 2009 by the pseudonymous Satoshi Nakamoto, and it remains the most widely known and used digital currency. However, Bitcoin's launch marked only the beginning of a much larger revolution in digital finance. Since Bitcoin, many other cryptocurrencies have been created, each with its unique features and use cases. One such cryptocurrency is **Ethereum**, which introduced a groundbreaking feature: **smart contracts**.

Smart contracts are self-executing agreements with the terms of the contract directly written into lines of code. These contracts automatically execute the agreed-upon terms when predefined conditions are met, without requiring intermediaries like lawyers or notaries. Smart contracts are a key driver of DeFi, enabling the automation of a wide range of financial services and applications on the blockchain.

DeFi is a movement that seeks to reconstruct and enhance traditional financial systems using blockchain technology, but without the need for centralized institutions such as banks, brokers, or insurance companies. Instead, DeFi applications rely on blockchain's transparency and security, powered by smart contracts to carry out transactions in a decentralized manner.

DeFi operates primarily on blockchain platforms. These networks support a range of financial services, including lending, borrowing, trading, savings, insurance, and even more sophisticated financial products such as derivatives and synthetics.

One of the key benefits of DeFi is its ability to provide financial services to people who have historically been excluded from the traditional financial system. Since DeFi platforms are decentralized and operate online, they allow anyone with internet access to use financial services, regardless of their location, background, or access to traditional banking.

Where traditional financial systems rely on trusted third parties – such as banks, brokers, and insurance companies – to facilitate transactions, DeFi empowers individuals to engage directly with each other. Transactions are governed by code rather than institutional authority, ensuring they are executed in a transparent, trustless, and secure manner. This removes many of the inefficiencies and risks inherent in traditional financial systems and hence offers greater yields and opportunities for individuals.

In short, cryptocurrency provides a new way to send and store value digitally, while DeFi builds on this foundation to create entire financial ecosystems that operate autonomously, without the need for centralized institutions. Together, they are forging a path to a more open, transparent, and decentralized financial world.

> "We have elected to put our money and faith in a mathematical framework that is free of politics and human error."
> Tyler Winklevoss (co-founder of Gemini)

THE TECHNOLOGICAL FOUNDATION OF CRYPTO AND DEFI

The technological foundation of both cryptocurrencies and DeFi is built upon blockchain technology. As we saw in the previous chapter, at its core, blockchain provides a revolutionary way to record and verify transactions in a decentralized and transparent manner, allowing digital assets and transactions to be securely managed without the need for centralized control. Instead of relying on a central authority, blockchain uses a distributed ledger system, where each participant in the network has access to the same data, ensuring that all transactions are visible and verifiable by everyone.

You will recall that blockchain technology creates a ledger of transactions that is immutable, meaning that once a transaction is recorded, it cannot be altered or deleted. This immutability and transparency provide an unparalleled level of security for users, reducing the risk of fraud or manipulation.

Blockchain operates through a system of consensus mechanisms, where the network participants must agree on the validity of each transaction before it is added to the ledger. This decentralized process, facilitated by cryptographic techniques, ensures that no single party can control or manipulate the network, providing trustless and autonomous financial systems.

When we talk about the technical architecture of blockchain, it is often described in terms of Layer 1 and Layer 2 solutions. Each plays a crucial role in the scalability and efficiency of the blockchain ecosystem.

LAYER I AND LAYER 2 BLOCKCHAINS

- **Layer 1** refers to the base-level blockchain infrastructure. This includes the core networks like **Bitcoin, Ethereum,** or other similar blockchains. Layer 1 blockchains are responsible for the essential operations such as network security, consensus (the process by which the network agrees on the validity of transactions), and data storage. These networks are the foundational elements that maintain the integrity and decentralization of the blockchain. However, as these networks grow in usage, they often experience scalability issues. During periods of high traffic, Layer 1 blockchains can become congested, resulting in slower transaction speeds and higher transaction fees.

  For instance, **Ethereum,** which is one of the most widely used blockchains, has often faced high gas fees when the network becomes congested with users trying to execute transactions or use decentralized applications (dApps).

- **Layer 2 solutions** are built on top of Layer 1 blockchains to address these scalability issues. These solutions do not require changes to the underlying blockchain itself but instead provide an additional layer that helps process transactions faster and more efficiently. Layer 2 solutions work by offloading some of the work from the main blockchain (Layer 1) to help improve transaction speeds and reduce costs.

  An example of a Layer 2 solution is **Polygon** for Ethereum. **Polygon helps scale Ethereum by processing transactions off-chain or on sidechains,** while still ensuring that they are recorded and validated securely on the Ethereum mainnet. This increases the transaction throughput, reduces costs, and enhances the user experience, all while maintaining security and decentralization, which are central to blockchain technology.

Layer 2 solutions play a critical role in enabling DeFi platforms to scale efficiently, as they ensure that DeFi applications, such as lending, borrowing, and trading, can operate smoothly and cost-effectively even during periods of high demand. These solutions also allow Ethereum and other Layer 1 blockchains to maintain their foundational principles of decentralization, transparency, and security, without compromising on performance. By using Layer 2s, DeFi platforms are able to offer a seamless experience to their users, making decentralized financial services more accessible and practical for everyday use.

As the Web3 ecosystem continues to grow and blockchain adoption accelerates, Layer 1 and Layer 2 solutions will continue to evolve, driving the expansion and development of decentralized financial systems across the globe. The combination of blockchain's transparency, immutability, and decentralization, paired with the scalability provided by Layer 2 solutions, holds the potential to transform how the world interacts with finance and digital assets.

SMART CONTRACTS: THE BACKBONE OF DEFI

Smart contracts are a cornerstone of DeFi and serve as the foundation for many of its key functions. These contracts are a form of self-executing agreement, where the terms and conditions are written directly into lines of code. Once the predefined conditions are met, the smart contract automatically executes the transaction or agreement without the need for human intervention or a trusted third-party intermediary. This feature is revolutionary because it eliminates the need for middlemen, significantly reducing transaction times and costs.

A smart contract's most powerful feature is its ability to automate processes in a trustless environment, where participants do not need to rely on a centralized authority to enforce the agreement. For example, in a DeFi lending protocol, when a borrower agrees to the terms of a loan, the smart contract can instantly transfer the required collateral – often in the form of cryptocurrencies like Ethereum – into the lender's wallet. Once the borrower repays the loan with interest, the smart contract automatically releases the collateral back to the borrower. This process is fully transparent, secure, and efficient.

In Figure 4.1 we can see that in TradFi for a payment you need an intermediary in a transaction, unlike with a smart contract where none are needed.

Similarly, smart contracts are also utilized in decentralized exchanges (DEXs) to facilitate peer-to-peer trading. In a DEX, like Uniswap, buyers and sellers are matched by smart contracts that execute trades automatically, removing the need for an exchange or platform to manage the trades. This creates a decentralized and automated marketplace where participants have full control over their funds.

Smart contracts have the ability to offer significant advantages over traditional financial systems. They remove intermediaries, reducing operational

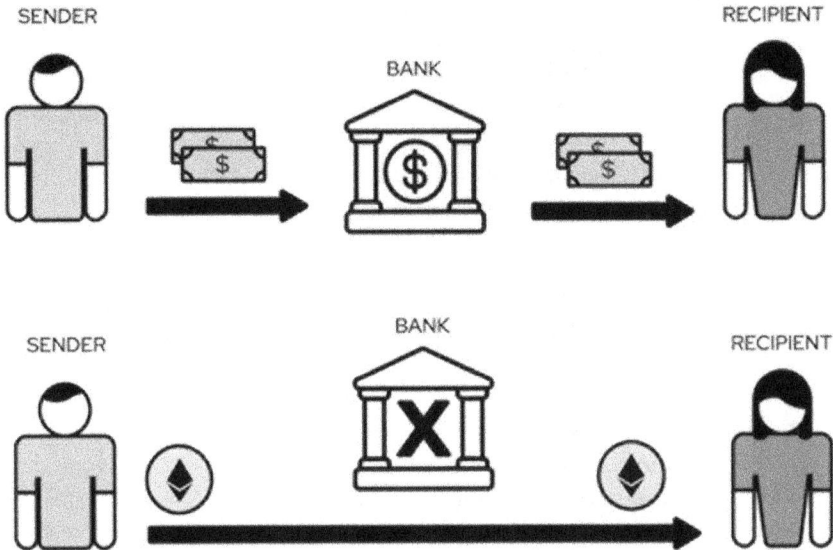

Figure 4.1 Cryptocurrency, such as Bitcoin, can pass from person to person without an intermediary involved.

costs and the possibility of human error or fraud. The transactions are recorded on the blockchain, making them immutable and transparent, which builds trust within the ecosystem. By automating the entire process, smart contracts also streamline operations, reducing the time it would take to complete complex financial transactions.

However, while smart contracts offer a number of advantages, they also come with inherent risks. Since they are written in code, the integrity of the contract is dependent on the code itself. If there are bugs, errors, or vulnerabilities in the code, it can lead to unintended consequences. For example, a poorly written smart contract could result in funds being locked or lost, or it could allow malicious actors to exploit a vulnerability, causing a potential security breach.

To mitigate these risks, it is crucial for developers to conduct rigorous code audits before deploying any smart contract on a blockchain. Auditing involves a thorough review of the code to ensure there are no vulnerabilities or errors that could be exploited. Additionally, smart contract developers must stay vigilant, keeping up with best practices for security and continually updating their contracts to address any emerging risks. Some DeFi projects also employ insurance protocols or decentralized governance models to protect users against potential losses due to smart contract failures.

In summary, smart contracts are a revolutionary aspect of DeFi that offer speed, efficiency, and decentralization. They eliminate the need for intermediaries, lower costs for users, and improve transparency for all. However, they are not without risks. Developers must take steps to ensure that smart contracts are secure and free from vulnerabilities in order to ensure the long-term viability and trustworthiness of DeFi platforms. As the DeFi space continues to evolve, smart contracts will remain a crucial building block, helping to drive the future of DeFi.

CASE STUDY: NEXUS MUTUAL – REVOLUTIONIZING INSURANCE WITH SMART CONTRACTS

A new era for insurance

Traditional insurance has always been a process laden with intermediaries – brokers, insurance companies, and regulatory bodies – that slow down the system and increase costs. But what if there was a way to eliminate all those middlemen, offer more transparency, and make the process faster and more secure? This is exactly the problem that Nexus Mutual set out to solve, and with its decentralized insurance platform powered by blockchain technology, it is reshaping the way we think about insurance.

Nexus Mutual operates on the Ethereum blockchain and uses smart contracts to completely rethink the insurance model. The platform empowers individuals to come together, pool risk, and purchase coverage – all without the need for traditional insurance companies. This decentralized structure offers a more transparent, fair, and accessible way for people to protect themselves from financial risk.

How nexus mutual works: The smart contract solution

The core innovation of Nexus Mutual lies in the use of **smart contracts** – self-executing contracts where the terms of the agreement are written directly into lines of code. This means that when certain conditions are met, the contract is executed automatically, removing the need for any intermediaries and ensuring that the process is quick and transparent.

For example, let's say someone wants to take out insurance for their cryptocurrency holdings. A smart contract would govern the policy, automatically processing claims without any human intervention. If a claim is made – for instance, a smart contract failure or a loss of funds in a hack – the smart contract is triggered and the payout is processed, all without the need for lengthy paperwork or approval processes.

Nexus Mutual takes this concept and applies it to a variety of insurance services, including smart contract failure coverage, business interruption insurance, and more. What makes this system stand out is its ability to offer **peer-to-peer** risk pooling, where the members themselves are responsible for underwriting and managing the coverage, rather than a centralized authority. This structure gives members more control over their policies and promotes transparency in how decisions are made.

The power of blockchain and decentralization

At the heart of Nexus Mutual is **blockchain technology**, which enables the decentralized and secure recording of all transactions. Every action – from premium payments to claims processing – is recorded transparently on the blockchain, ensuring that no one can alter or erase any data once it's been written. This is a significant departure from traditional insurance, where processes are often opaque and subject to manipulation.

The use of blockchain provides an inherent level of **security** and **trust**. Since the entire process is decentralized, there's no single point of failure, and all claims are validated by a collective, rather than an individual authority. This decentralized approach helps to reduce fraud, increase efficiency, and ultimately deliver better outcomes for participants.

Decentralized governance: The nexus mutual model

One of the most innovative aspects of Nexus Mutual is its **governance model**. Instead of being run by a centralized insurance company, the platform is governed by its members through **tokenized governance**. Members purchase Nexus Mutual's governance tokens (NXM), which give them the ability to vote on key decisions, such as claims assessments and policy terms.

This **community-driven model** helps ensure that the platform operates in a way that benefits all participants. For example, when a claim is made, it is assessed by the community and decided upon by a group of token holders. This reduces the risk of unfair decisions, as the process is transparent and open to scrutiny. It also helps lower the costs typically associated with insurance, by removing the need for expensive intermediaries.

The roadblocks and future potential

Despite its promising start, Nexus Mutual does face challenges – most notably in the areas of **regulatory uncertainty** and **smart contract vulnerabilities**. As a decentralized platform, it doesn't fall under traditional insurance regulations, which has raised concerns among governments and regulators about its compliance with existing laws. As DeFi and blockchain sectors mature, the regulatory landscape will likely evolve, but this is still an area where Nexus Mutual must navigate carefully.

Furthermore, smart contracts, while revolutionary, are not foolproof. The code that governs these contracts must be thoroughly tested to avoid bugs or vulnerabilities that could lead to unintended consequences, such as lost funds or the failure of a payout. Nexus Mutual has addressed this challenge by conducting rigorous audits of its contracts and encouraging transparency across the platform, but it remains a risk that all blockchain-based applications must manage.

Despite these challenges, the potential for Nexus Mutual and similar platforms is enormous. As more people become aware of DeFi and blockchain technologies, the opportunity to provide low-cost, transparent, and efficient insurance will only grow. The ability for people worldwide to access insurance without the need for traditional financial institutions is a game changer, especially for underserved populations who have historically been excluded from the traditional insurance system.

> "In the decentralized future, finance will be built by code and run on the blockchain."
> Ryan Sean Adams (founder of Bankless)

COMPARING TRADFI WITH CRYPTO AND DEFI

The traditional financial system, commonly referred to as **TradFi**, is built on centralized institutions like **banks, payment processors,** and **insurance companies.** These organizations serve as intermediaries, facilitating financial transactions, managing assets, and providing various services such as loans, savings accounts, and insurance. While these systems have long been integral to the global economy, they come with several significant drawbacks.

One of the most glaring issues with TradFi is the **cost.** Fees in TradFi are often high, particularly for cross-border transactions or even basic services like withdrawals or fund transfers. Additionally, these systems are prone to **delays,** with transactions taking days, especially when dealing with international transfers. This is because multiple intermediaries, such as clearinghouses, banks, or international payment networks, need to be involved before a transaction is complete. This adds inefficiency, particularly for people needing faster transactions or for businesses working with global clients.

Beyond fees and delays, there's also the issue of **trust.** Financial institutions are built on the assumption that they will be trustworthy and reliable. However, this trust can be breached, as seen in the aftermath of financial crises, fraud, or mismanagement by large institutions. In some cases, errors or fraud by these intermediaries have led to significant financial losses. Equally, the fact of having an intermediary involved in the transaction can also lead to that intermediary blocking transactions or withholding funds.

In stark contrast, **cryptocurrency** and **DeFi** eliminate intermediaries altogether. By relying on **blockchain** technology, these innovations offer an alternative that is not only more **efficient** and **cost-effective**, but also **secure** and **transparent**.

In the crypto and DeFi ecosystem, the power is shifted away from centralized authorities, empowering individuals with the ability to manage their own finances. The key feature of these technologies is their **decentralized** nature. Anyone with an internet connection can participate in the system, allowing access to financial services regardless of location, wealth, or background. This is particularly valuable in underserved areas, where access to traditional banking is limited or unavailable.

With **crypto** like Bitcoin or Ethereum, transactions are **peer-to-peer**, meaning there's no need for a central authority to facilitate or process the transactions. Instead, transactions are verified and recorded by a **distributed network** of miners or validators, making the process far more transparent and secure. This decentralization eliminates the middlemen that typically charge fees, delay transactions, and potentially introduce fraud.

REAL-WORLD EXAMPLES OF TRADFI VS. DEFI

Example 1: Transferring funds

In TradFi, transferring money between different financial institutions, especially for international payments, can be a drawn-out process. If you need to send money from one country to another, you're likely to go through intermediaries like **SWIFT**, which ensure that the transaction is verified, funds are moved, and accounts are updated. However, these steps can take several **business days**, especially for international transfers. On top of that, you'll likely face hefty **fees** for the service, which can add up significantly depending on the size of the transfer.

In comparison, cryptocurrencies like **Bitcoin** offer a drastically different solution. With Bitcoin, you can transfer funds directly from your wallet to anyone around the world. There's no need for an intermediary, and the transaction happens **on the blockchain**, where it is recorded transparently. These transactions typically take only **minutes** or **seconds** to complete and cost a fraction of the fees traditionally charged by banks or payment processors. Additionally, the transaction is **immutable**, meaning once it's on the blockchain, it cannot be altered, offering a level of security and transparency far beyond traditional systems.

Example 2: Trading assets

The world of trading is another area where DeFi is providing a clear alternative to traditional financial systems. Traditionally, if you wanted to trade stocks or other assets, you would use a **broker** or an **exchange**. These platforms match buyers with sellers, charging fees for facilitating the trades and often controlling the process in ways that benefit the institution itself. The centralized nature of these exchanges also means that they hold all the control over the assets being traded.

On the other hand, **DEXs** like **Uniswap** or **Jupiter** offer a radically different approach. These platforms use **smart contracts** and **liquidity pools** to allow users to trade **directly** with one another, without relying on a central exchange or intermediary. By leveraging **blockchain technology**, DEXs can offer users **more control** over their trades, lower fees, and better **privacy**. Liquidity providers on DEXs often earn rewards for contributing to the liquidity pools, further incentivizing participation and promoting decentralization.

While traditional financial systems still play a crucial role in the global economy, **crypto** and **DeFi** are quickly gaining ground as viable, alternative solutions. By removing intermediaries, reducing costs, and increasing speed, transparency, and security, these technologies are fundamentally reshaping the financial landscape.

In the coming years, as the technology matures and adoption grows, it is likely that DeFi will continue to challenge the **status quo**. Whether it's for transferring funds or trading assets, the shift toward **peer-to-peer** systems offers significant benefits that could change how people interact with money and financial services across the globe. The future of finance may very well be decentralized, and the technology behind crypto and DeFi is the key to unlocking that future.

> *"Fundamentally what blockchain lets you do is, it lets you write code that can make strong commitments about how it will behave in the future."*
> Chris Dixon (general partner at Andreessen Horowitz)

THE KEY PRINCIPLES OF CRYPTO AND DEFI

Building upon what we have just seen about the future of finance likely being decentralized – we should pause and understand what that really means in detail. The rise of cryptocurrencies and DeFi is founded upon several key principles that not only distinguish them from traditional financial systems, but also hold the potential to dramatically reshape how we engage with money, assets, and financial services. Let's dive into these principles in more detail.

- **Decentralization:** Traditional financial systems have been largely dominated by centralized entities, such as banks, payment processors, and insurance companies. These institutions are the gatekeepers of financial services, holding the power to facilitate transactions, manage assets, and set terms of service. The inherent challenge here is that these systems are controlled by a small group of centralized entities, which introduces inefficiencies, risks, and opacity into the financial system.

  In contrast, cryptocurrencies and DeFi operate on decentralized networks, where there is no single entity with complete control. Instead, decisions are made collectively through consensus protocols, allowing all participants to have a say in the governance and functioning of the network. This decentralization enhances transparency, as the rules governing the network are publicly accessible, and it significantly reduces the chances of a single point of failure.

  Moreover, decentralized systems are inherently more resilient to disruptions, whether due to economic downturns, natural disasters, or even systemic fraud. The removal of middlemen means lower costs and increased efficiency, empowering individuals to engage in financial activities without relying on traditional gatekeepers.

- **Trustlessness:** At the heart of DeFi lies the concept of trustlessness – an idea that completely changes how we think about relationships in financial transactions. In TradFi, trust is placed in intermediaries like banks, brokers, and financial institutions, whose roles are based on their reputation, regulatory oversight, and business practices. This reliance on trusted third parties introduces the risk of fraud, manipulation, and delays. Furthermore, the human element can lead to errors or biases in decision-making.

  In the DeFi world, however, trust is no longer placed in individuals or institutions, but in the underlying code. Smart contracts, which are self-executing agreements, automatically enforce terms when certain conditions are met, and their execution is irreversible and transparent. This makes it possible for two parties to transact directly with one another without needing to trust a central authority.

  For example, in a lending protocol, once the borrower submits collateral and agrees to the terms, the smart contract will ensure the loan terms are adhered to – removing the risk of human error or dishonesty. The transparency of the blockchain ensures that all transactions are visible and auditable by anyone, further reducing the potential for fraud.

- **Financial inclusion:** One of the most transformative aspects of DeFi is its potential to bring financial services to those who have historically been excluded from traditional banking systems. Globally, millions of people remain unbanked or underbanked, often due to geographical barriers, lack of access to physical bank branches, or the high costs of traditional banking services. DeFi eliminates these barriers by enabling anyone with an internet connection to access a full range of financial services, from lending and borrowing to trading and investing.

  DeFi platforms function without the need for intermediaries like banks, enabling peer-to-peer transactions and reducing the costs associated with traditional financial services. For example, someone in a rural area can lend or borrow money through a decentralized platform, without needing to go through a bank or credit union. Similarly, those in developing regions can access investment opportunities that were previously limited to wealthy individuals or institutions. This inclusivity promotes economic empowerment, allowing individuals around the world to take part in the global financial ecosystem, regardless of their location or financial status.

- **Programmable money:** Another crucial principle of cryptocurrencies and DeFi is their programmability. Unlike traditional currencies, which are relatively static in their function, cryptocurrencies and digital tokens can be programmed to serve specific purposes within the DeFi ecosystem. This programmability allows for greater flexibility and customization, enabling new financial products and services that traditional financial systems cannot offer.

  For example, stablecoins such as USDT and USDC are designed to maintain a stable value by being pegged to real-world assets like the US dollar. These coins are widely used within the DeFi space to reduce the volatility often associated with cryptocurrencies. Programmable money also includes governance tokens, which give holders a voice in the decision-making processes of DeFi platforms. Users can vote on proposals, such as platform upgrades or changes in the system, allowing for decentralized, community-driven governance.

  This flexibility enables a wide variety of financial use cases, from yield farming to decentralized insurance, where each token can be customized to meet the needs of the platform and its users. The ability to create money that can be adjusted for specific use cases opens up a whole new range of possibilities for innovation, creating financial tools that are more aligned with the needs of the user rather than the centralized authority.

Given that these key principles and technology are now being utilized, let us turn our attention to how they are being implemented today.

CURRENT USE CASES FOR CRYPTO AND DEFI

The DeFi ecosystem has experienced rapid growth in recent years, introducing an array of innovative applications. These solutions are changing how we interact with money, banking, and finance by offering alternatives that are faster, more secure, and accessible to everyone. Below are some of the most popular and impactful use cases in the DeFi space.

1. **Stablecoins**
   Stablecoins are digital currencies that aim to minimize the volatility typically associated with cryptocurrencies like Bitcoin or Ethereum by pegging their value to traditional assets, such as the US dollar or gold. These stable values offer the benefits of cryptocurrencies – fast transactions, global accessibility, and decentralized control – while mitigating the price swings that can make crypto assets difficult to use in everyday transactions. The adoption of stablecoins such as USDT (Tether), DAI, and USDC has surged in DeFi applications, making them a critical tool in decentralized lending, borrowing, and trading. Their stable nature makes them ideal for maintaining a consistent store of value within the DeFi ecosystem.

2. **Decentralized lending and borrowing**
   DeFi platforms such as Aave, Compound, and MakerDAO offer a transformative approach to lending and borrowing by removing traditional financial institutions from the equation. These platforms operate through smart contracts, allowing users to lend their cryptocurrency assets in exchange for interest or borrow against their holdings without the need for intermediaries like banks. The process is decentralized, meaning that users interact directly with the platform, where terms are automatically executed based on smart contract code. This not only reduces costs but also democratizes access to financial services, offering opportunities to individuals who may not have access to traditional banking systems.

3. **DEXs**
   DEXs, such as Uniswap, SushiSwap, Jupiter and Curve Finance, allow users to trade cryptocurrencies directly with one another without relying on centralized exchanges like Coinbase or Binance. These exchanges operate based on automated market makers (AMMs), which enable transactions between buyers and sellers through liquidity pools rather than order books. The use of AMMs makes trading more efficient and decentralized, eliminating the need for intermediaries and often providing lower fees. DEXs empower users by giving them full control of their funds, reducing the risks associated with centralized platforms where assets may be vulnerable to hacks or mismanagement.

4.   **Yield farming and staking**

Yield farming and staking are two increasingly popular strategies in the DeFi space for earning passive income from crypto assets. Yield farming involves providing liquidity to decentralized platforms, such as lending pools or exchanges, with the promise of rewards, typically in the form of platform tokens. These rewards can accumulate over time, making yield farming a way for crypto holders to generate income while maintaining their exposure to cryptocurrency markets. Similarly, staking allows users to lock up their crypto assets in a network to help secure it, such as through proof-of-stake (PoS) consensus mechanisms. In exchange for staking their assets, participants earn rewards. Both yield farming and staking offer compelling incentives for those looking to earn additional returns on their crypto holdings without selling them.

5.   **Decentralized autonomous organizations (DAOs)**

DAOs represent a groundbreaking shift in how organizations are governed. We will investigate them in a lot more depth later on. Unlike traditional organizations with hierarchical structures and centralized leadership, DAOs operate without a central authority. Instead, decision-making is distributed across a community of stakeholders, each of whom can influence the direction of the organization through the use of governance tokens. These tokens allow participants to vote on proposals, whether for changes to the protocol, funding decisions, or other operational matters. DAOs use smart contracts to execute decisions transparently and efficiently. One of the most notable examples is MakerDAO, which governs the DAI stablecoin, ensuring that the platform remains decentralized and aligned with its users' interests. DAOs are paving the way for more inclusive, transparent, and democratic forms of governance within the blockchain ecosystem.

These use cases show just a glimpse of the immense potential that DeFi offers. As the space continues to evolve, more innovative applications are likely to emerge, further disrupting traditional financial systems and unlocking greater opportunities for individuals around the world.

THE FUTURE OF CRYPTO AND DEFI

The potential of cryptocurrency and DeFi is vast, and we are just scratching the surface of what can be achieved. Blockchain technology, which underpins both crypto and DeFi, is continually evolving, with new innovations on the horizon. The future of these technologies promises to reshape the financial landscape, offering enhanced opportunities for both individuals and businesses. Let's explore what we might expect in the coming years.

- **Integration with TradFi:** As DeFi matures and becomes more widely adopted, there is a growing possibility that it will integrate with TradFi. The financial services sector, long dominated by established institutions such as banks and insurers, is seeing increasing interest in blockchain-based solutions. Many traditional financial entities are now exploring how to incorporate DeFi products into their services, which could lead to a hybrid financial ecosystem.

  The convergence of DeFi and TradFi would allow both systems to coexist, bringing the benefits of decentralization to established financial products while retaining the reliability and stability of traditional financial institutions. For example, decentralized lending platforms might partner with banks to offer interest-bearing accounts or loans that are governed by blockchain technology, increasing the accessibility and efficiency of these services. This hybrid approach would expand financial services and open new possibilities for users and financial institutions alike.

- **Increased regulation:** As the DeFi space continues to grow, regulatory bodies are beginning to take a more active role in shaping its development. One of the main challenges for DeFi has been the lack of clear regulations, which has created uncertainty for both developers and users. While some argue that regulation could stifle innovation, others believe that establishing clear rules will help protect consumers and reduce fraudulent activities in the ecosystem.

  Governments and regulators will likely continue to explore ways to bring DeFi under a regulatory framework. This could include guidelines for anti-money laundering (AML) and know your customer (KYC) compliance, as well as protections for users participating in decentralized platforms. As the industry matures and more mainstream users become involved, we can expect to see increased regulatory oversight, ensuring that DeFi platforms remain secure, transparent, and trustworthy.

- **Privacy and security:** Privacy and security are paramount in any financial system, and as the DeFi ecosystem grows, so too will the need for robust security solutions. Blockchain technology is inherently secure due to its decentralized nature, but it is still vulnerable to hacks, smart contract vulnerabilities, and other attacks. All of which are points of concern developers within the Web3 ecosystem are actively working to remedy.

  With regards to privacy, innovations such as zero-knowledge proofs (ZKPs) are likely to play a crucial role; they allow for the verification of transactions without revealing the data involved in the transaction itself, such as the identities or amounts involved

in a transfer. These privacy-enhancing technologies will help protect users' personal information and ensure that their transactions
remain secure, even in a transparent and public ledger system like
blockchain. The development of these privacy solutions will be critical in attracting more institutional investors and everyday users to
the DeFi space.

"The future of money is digital."
Christine Lagarde (president of the European Central Bank)

Looking ahead, the future of cryptocurrency and DeFi is incredibly exciting.
As blockchain technology continues to mature, we can expect to see more
advanced financial products that offer new ways to invest, borrow, lend,
and transact. With improved interoperability, DeFi platforms will be able
to communicate seamlessly with traditional financial systems, offering users
more choices and access to financial services globally. In the coming years,
we may witness the development of hybrid financial systems, increased
regulatory clarity, and enhanced privacy and security measures, all of which
will help DeFi reach its full potential. The pace of innovation will continue
to accelerate, making crypto and DeFi integral parts of the global financial
ecosystem.

ACTIONABLE STEPS : HOW TO GET STARTED WITH CRYPTO AND DEFI

Starting your journey into the world of cryptocurrency and DeFi can be incredibly rewarding, but it's important to take a structured approach. Here's how
to break it down into manageable steps, so you can build your understanding,
get hands-on experience, and start navigating the DeFi ecosystem.

1. Start by learning about cryptocurrencies and DeFi

Before diving in, it's crucial to familiarize yourself with the foundational
concepts. Understanding how cryptocurrencies work and what DeFi is all
about will give you a solid base to build on.

- **Online courses and resources**: Take advantage of free learning
  platforms. Websites like **Coinbase Learn**, **Binance Academy**, and
  Ethereum.org offer excellent resources for beginners. They provide

a comprehensive understanding of the core concepts of blockchain, crypto, DeFi protocols, and smart contracts.

- **Books and podcasts**: This book of course gives you a great base, but consider reading more books like *The Basics of Bitcoins and Blockchains* by Antony Lewis for deeper insights. Many podcasts also offer real-time updates and discussions about the DeFi ecosystem.
- **Community engagement**: Engage in crypto and DeFi communities online. Join platforms like Reddit (r/Cryptocurrency and r/DeFi), X (formerly Twitter), and Telegram to connect with others and stay updated on trends and new developments.

2. Try yield farming or staking

Using your digital wallet you can get involved with one of the easiest ways to start earning passive income with your crypto through yield farming or staking. Both methods allow you to put your crypto to work while gaining interest or rewards.

- **Yield farming**: Yield farming involves lending your crypto to DeFi protocols or liquidity pools in exchange for rewards, often paid in the platform's native token. You can explore platforms like **Yearn.Finance**, **MarginFi**, **Aave**, or **Compound** for yield farming opportunities. These platforms offer various ways to earn returns on your cryptocurrency assets from the lending to liquidity provision.
- **Staking**: Staking involves locking up your crypto to help secure a blockchain network or dApp. In return, you earn rewards. For example, with **Ethereum 2.0**, you can stake ETH and earn staking rewards. Staking is less risky than yield farming but typically offers lower returns.

3. Experiment with DEXs

DEXs allow you to trade crypto directly with other users, eliminating the need for a centralized third party like Coinbase or Binance.

- **Start trading**: Platforms like **Uniswap**, **SushiSwap**, **Jupiter** and **Curve Finance** are popular DEXs that allow you to swap tokens directly with others. These exchanges use AMMs to match buyers and sellers.
- **Provide liquidity**: On DEXs, you can also provide liquidity to trading pairs in exchange for a share of the fees generated by the platform. For example, if you provide liquidity for a USDT/ETH pair on Uniswap, you earn a portion of the transaction fees.
- **Learn about AMMs**: AMMs are the heart of DEXs. They remove the need for a traditional order book and facilitate peer-to-peer trading by automatically setting prices based on the liquidity pool and algorithm.

Familiarizing yourself with how they work will make your trading experience much more effective.

4. Stay informed and keep learning

The DeFi space is rapidly evolving, and new developments are occurring daily. It's crucial to stay informed about the latest updates and trends.

- **Follow industry news**: Subscribe to newsletters like *The Defiant*, *DeFi Pulse*, and *CoinDesk* to receive updates and analysis on the latest developments in DeFi. Keeping an eye on emerging trends and new protocols will give you a competitive edge.
- **Join online communities**: Engage with other users in DeFi-focused forums like **DeFi Llama**, **Discord** channels, and **Telegram groups**. These communities are invaluable for asking questions, sharing experiences, and learning from others who have been in the space longer.
- **Experiment and test**: Start small by experimenting with different platforms and tools. Test different strategies in yield farming, staking, and trading to better understand the risks and rewards. This hands-on experience will be invaluable in making informed decisions as you expand your involvement in DeFi.

By following these actionable steps, you'll have a comprehensive understanding of how to navigate the world of crypto and DeFi. Whether you want to earn passive income, trade on DEXs, or contribute to governance through DAOs, these steps provide the foundation to help you confidently enter the DeFi ecosystem.

Chapter 5

NFTs decoded

Transforming digital ownership

Non-fungible tokens (NFTs) have captured the world's attention, transforming everything from art to music, gaming to real estate, and even intellectual property (IP). But what exactly are NFTs? How do they work, and why are they important? In this chapter, we will explore these topics in depth and examine the world of NFTs, explaining their significance, their evolving use cases, and the potential they hold for industries and consumers alike.

INTRODUCING NFTS: THE CONCEPT OF NON-FUNGIBILITY

To truly understand NFTs, we need to begin with the concept of **fungibility**. Fungible assets are those that are interchangeable because each unit is identical in value. **Fiat currency** (such as dollars, euros, and yen) is a prime example – if you exchange a $20 note for another $20 note, the value remains the same, and there is no differentiation between the two.

On the other hand, **non-fungible assets** are unique and cannot be replaced by another identical item. Think of a concert ticket for a VIP seat – it cannot be swapped one-to-one for a standard admission ticket, as they offer different experiences as shown in Figure 5.1. This is what makes NFTs so distinctive in the digital world.

NFTs are digital representations of unique, scarce assets. They can represent anything that is considered valuable due to its uniqueness – whether that's digital art, a rare collectible, or even a piece of virtual real estate in a metaverse. Unlike cryptocurrencies such as Bitcoin or Ethereum, which are fungible and interchangeable, each NFT is one of a kind, and its ownership is recorded on a blockchain, guaranteeing its authenticity and uniqueness.

DOI: 10.1201/9781003616504-5

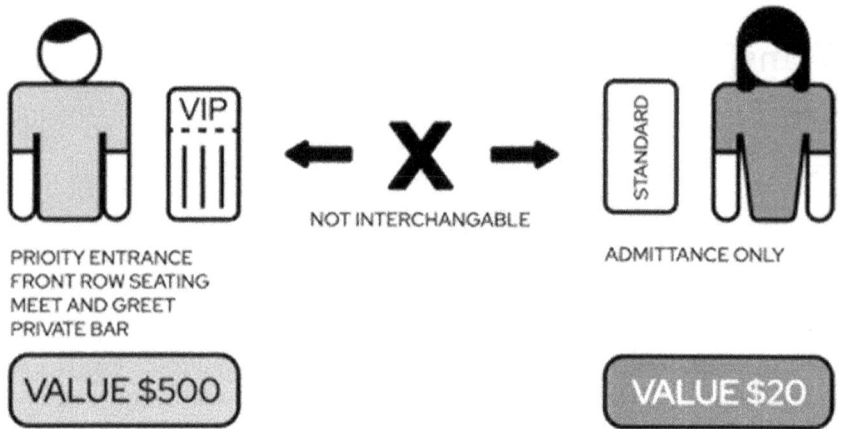

Figure 5.1 Tickets that offer different experiences (in this case a $500 ticket and a $50 ticket) cannot be swapped one-to-one as they are non-fungible, unlike a $20 note that is fungible and consistent in its nature.

THE EVOLUTION OF NFTS: FROM COLLECTIBLES TO INNOVATION

NFTs began as a way to buy and sell digital collectibles. One of the earliest and most notable examples was **CryptoKitties,** a blockchain-based game that allowed players to buy, sell, and breed digital cats. While the CryptoKitties project started as a fun experiment, it introduced the world to the idea of owning and trading unique digital items, laying the groundwork for the broader NFT ecosystem.

Since then, the world of NFTs has evolved rapidly. The rise of **digital art** has been a major factor in NFTs' mainstream success. One of the landmark events in the NFT space was the sale of **Beeple's 'Everydays: The First 5000 Days'** artwork at Christie's auction house for a staggering $69 million in 2021. This sale proved that NFTs could represent high-value digital assets with significant cultural cachet.

However, the evolution of NFTs extends far beyond just digital art. NFTs are now being used to represent a wide variety of digital and physical assets, including **virtual real estate, in-game items, domain names,** and even **IP.** Major brands and industries have started to integrate NFTs into their business models, exploring new ways to engage consumers and create value. The rise of the **metaverse** has further expanded the potential uses for NFTs, as virtual worlds create opportunities for users to buy, sell, and trade digital land, avatars, and assets.

> *"NFTs... and then it's on the blockchain and everybody sort of agrees, whatever you just made is part of the official blockchain. And it is tokenized then. And you can sell that token or you can give somebody that token or whatever."*
> Mike Winkelmann aka Beeple (digital artist)

HOW DO NFTS WORK?

At the core of NFTs lies **blockchain technology** – the same technology that powers cryptocurrencies like Bitcoin and Ethereum. Blockchains are decentralized, transparent ledgers that record transactions across a distributed network of computers. This decentralized nature is what makes NFTs so secure, as no single entity has control over the data.

NFTs are typically minted on blockchain platforms like **Ethereum**, which is the most widely used blockchain for NFTs. When an NFT is created, or 'minted,' it is assigned a unique identifier, which is recorded on the blockchain. This identifier contains information about the asset's ownership, its history, and its characteristics – data that cannot be altered once the NFT has been minted.

NFTs are created using **smart contracts**, which are self-executing contracts with the terms written directly into code. These smart contracts automate processes like the transfer of ownership and the distribution of royalties, making NFT transactions seamless and efficient. For example, when an NFT is resold, the smart contract can automatically ensure that the original creator receives a percentage of the sale, providing a new revenue stream for artists and content creators.

Each NFT is composed of **metadata** – this includes key information about the asset, such as its title, description, and even media files like images, videos, or audio. Metadata is typically stored off-chain, but the blockchain ensures that the token's ownership and transaction history remain immutable and transparent.

THE PURPOSE OF NFTS: VERIFIABLE OWNERSHIP AND DIGITAL SCARCITY

NFTs serve as a solution to a key problem in the digital world – **ownership and scarcity**. While digital assets can be easily replicated, the notion of ownership is often unclear. You can copy a digital image or video as many times as you like, but only one person can truly own the original.

NFTs provide **verifiable ownership**. When someone purchases an NFT, they are not just buying a file; they are purchasing ownership of the original digital asset, as verified by the blockchain. The blockchain ledger provides an irrefutable proof of ownership that can't be tampered with, ensuring that the asset is authentic and unique.

NFTs also solve the issue of digital scarcity. Unlike physical objects, digital assets can be reproduced endlessly. However, by turning an asset into an NFT, creators can impose **scarcity** – they can limit the number of NFTs representing a particular piece of content, whether it's a piece of art, a concert ticket, or even a piece of virtual real estate. This scarcity, combined with verifiable ownership, drives the value of NFTs.

THE BENEFITS OF NFTS: EMPOWERING CREATORS AND CONSUMERS

The ownership and scarcity paradigm has meant that NFTs can offer a wide range of benefits to both creators and consumers. For **creators**, NFTs provide a way to monetize digital content in new ways. Artists, musicians, and other content creators can mint their work as NFTs and sell it directly to their audience, bypassing traditional intermediaries like galleries, record labels, or streaming platforms.

One of the most powerful features of NFTs is the **royalty mechanism**. When an NFT is resold, the original creator can receive a percentage of the sale automatically, thanks to smart contracts. This creates a **continuous revenue stream** for creators, ensuring that they benefit from the increasing value of their work as it is resold on the secondary market.

For **consumers**, NFTs provide a way to own unique digital assets and engage with their favorite creators. NFTs can represent a wide range of products, from digital art to in-game items, and even access to exclusive experiences. By owning an NFT, consumers can gain access to a growing ecosystem of digital and physical benefits, including community engagement, event access, and exclusive content.

REAL-WORLD EXAMPLES: HOW BRANDS ARE USING NFTS

NFTs have caught the attention of brands across various industries, with many leveraging them as a tool for **consumer engagement** and **brand loyalty**.

Case Study 1: Starbucks Odyssey – Revolutionizing customer engagement with NFTs

Starbucks, long known for its innovative customer loyalty programs, took a bold step into the world of **NFTs** in 2022 with the launch of its **Starbucks**

Odyssey program. Starbucks Odyssey is a **Web3 rewards program** that allows customers to collect and trade **unique NFT stamps** as part of their loyalty experience. Each NFT stamp offers exclusive access to rewards, experiences, and even Starbucks merchandise, providing customers with an exciting way to deepen their engagement with the brand.

What makes Starbucks Odyssey particularly unique is its integration of **blockchain technology** through **Polygon**, a more energy-efficient blockchain. This choice ensures that customers can interact with NFTs in a sustainable and cost-effective manner, addressing concerns around the environmental impact that has been associated with some blockchain networks. By leveraging **Polygon**, Starbucks enables a smoother, eco-friendly interaction with NFTs, positioning the brand as a responsible leader in digital innovation.

In terms of business strategy, Starbucks Odyssey has proven to be a resounding success, and as they concluded this test project in 2024 they will firmly have eyes on the future. Not only has it been an opportunity to tap into the growing demand for **Web3 technologies**, but it has also **generated significant revenue** through the sale of NFTs. This initiative has enabled Starbucks to maintain its position at the forefront of digital transformation, blending traditional customer loyalty with the latest advancements in blockchain and cryptocurrency.

Customers are not just rewarded for buying coffee – they are also incentivized to **collect NFTs**, opening up opportunities for **exclusive digital experiences** and even events that would otherwise be out of reach. This integration of digital assets with real-world rewards has generated a new kind of **brand loyalty**, with customers feeling more connected to the brand's community.

Case Study 2: Red Bull Racing – Integrating NFTs into motorsports for enhanced fan engagement

Red Bull Racing, a powerhouse in **Formula 1**, has leveraged the power of NFTs to redefine fan engagement in the high-speed world of motorsports. The team launched a series of **limited-edition NFTs** that feature **digital versions** of their iconic **race cars** and **drivers**, which are sold during major **Formula 1 events**.

These NFTs serve as **digital memorabilia**, offering fans a new way to commemorate their passion for the sport and their loyalty to the team. Red Bull Racing's NFT auctions, often taking place during live races, have created a **real-time connection** between the team and its fan base, making the experience more immersive and exclusive. Fans can own a piece of **Formula 1 history** through these unique digital assets, which is more than just a keepsake – it's a digital experience that ties them to the team.

What sets Red Bull Racing's NFT strategy apart is the seamless integration of **digital** and **physical** experiences. These NFTs aren't just static images; they offer opportunities for fans to unlock **virtual racing simulators**, interact with the team through exclusive **behind-the-scenes content**, and enjoy **virtual meet-and-greets** with drivers. By creating these interactive layers, Red Bull Racing has successfully blurred the line between the **digital and physical worlds**.

The integration of NFTs into motorsports creates a **new revenue stream** for Red Bull Racing, providing fans with **novel ways to connect** with the team beyond the traditional sponsorships and ticket sales. This innovative use of NFTs is setting a new benchmark for **sports teams** looking to explore the potential of Web3 technologies. Red Bull Racing's ability to blend fan engagement, exclusivity, and technology into a unified strategy has proven to be an innovative approach to sports marketing.

Both of these examples highlight how **NFTs** are not just digital trends but powerful tools for engaging and expanding communities. From coffee cups to race cars, NFTs are reshaping the way brands and fans interact, offering exclusive rewards, new revenue streams, and immersive experiences that connect the digital world with the real world in unprecedented ways.

GETTING INVOLVED WITH NFTS: HOW TO ENTER THE WORLD OF DIGITAL OWNERSHIP

As we've just seen, NFTs offer a unique way to own, trade, and collect digital assets, but the process of getting involved might seem a little daunting at first. Fortunately, entering the world of NFTs has never been easier. The ecosystem is growing rapidly, and there are now a variety of platforms and tools that can help creators, collectors, and investors navigate this exciting space.

CREATE AND MINT NFTS

If you are a digital artist, musician, or game developer, one of the first steps in entering the NFT world is creating and minting your own NFTs. **Minting** refers to the process of turning your digital creation into a unique, tradeable NFT by recording it on the blockchain.

Several platforms have made this process accessible including:

- **OpenSea:** One of the most popular NFT marketplaces, OpenSea offers a seamless minting process where you can create your NFTs, list them for sale, and track their sales history. It also integrates well with multiple wallets, allowing for easier transactions.
- **Rarible:** Another easy-to-use platform, Rarible allows users to mint NFTs without needing to be a blockchain expert. It's an open platform where artists can set up their own collections and start selling digital works directly to buyers.
- **Mintable:** Mintable makes the NFT creation process simple with its intuitive design and various customization options. Artists and creators can use it to mint and sell digital artwork, music, and other collectibles.
- **Art Blocks:** Art Blocks specializes in generative art, allowing artists to create and mint NFTs that are algorithmically generated based on specific code. It's a platform where art and technology intersect, and it focuses on high-quality, programmable artwork. Artists can showcase and sell unique generative art pieces, and collectors can explore and purchase them directly from the platform.
- **fxhash:** fxhash is a marketplace for generative art NFTs on the Tezos blockchain. This platform is community-driven and offers creators the ability to mint generative art pieces in an easy and accessible way. The minting process on fxhash is simple, and it provides artists with the opportunity to connect with a unique community of collectors and other generative artists.

These platforms typically offer a user-friendly interface, even for beginners, allowing creators to easily upload digital works (images, music, videos, etc.), set up smart contract terms, and list their NFTs for sale in various markets. Minting can often be done with just a few clicks, and the process is designed to be accessible to all levels of creators, from newcomers to experienced artists.

PARTICIPATE IN NFT MARKETPLACES

Once you've minted an NFT, the next step is to explore the world of buying and selling these digital assets. **NFT marketplaces** are platforms that allow users to trade NFTs, and many of them offer diverse categories, from art and collectibles to virtual land and gaming assets. Some of the most well-known NFT marketplaces include:

- **OpenSea:** One of the largest and most popular NFT marketplaces, OpenSea has a vast collection of assets, from digital art to rare collectibles. It offers search and filter tools to help users discover new NFTs based on categories, such as art, virtual worlds, and sports.

- **SuperRare:** A high-end NFT marketplace that focuses on curated digital artwork, SuperRare is perfect for collectors and investors who are looking for unique, limited-edition pieces. It provides a platform for artists to showcase their work and for collectors to buy rare, one-of-a-kind pieces.
- **Art Blocks:** Besides being a minting platform, Art Blocks also functions as a marketplace. Once generative art is minted, it is listed on the platform's marketplace for collectors to purchase. Art Blocks is known for its high-quality art and well-curated collections.
- **fxhash:** As mentioned, fxhash is not just a place to mint generative art; it's also a vibrant marketplace where collectors can purchase unique, algorithmically generated art pieces. The platform connects creators with collectors, and its focus on the Tezos blockchain gives it a distinct niche in the NFT world.

These marketplaces offer easy access to buying and selling NFTs, and many of them allow for diverse payment methods, including cryptocurrency or fiat currencies. By using these platforms, collectors can discover hidden gems, interact with artists, and build up their own NFT portfolios.

JOIN NFT COMMUNITIES

In the world of NFTs, the **community** is a key driver of innovation, and engaging with other creators, collectors, and enthusiasts can help you stay informed and find new opportunities. There are many communities across social media platforms where you can connect, learn, and share your NFT journey:

- **X** (formerly Twitter): The go-to platform for NFT news, discussions, and updates. Many NFT creators, influencers, and enthusiasts use X to post updates, share trends, and promote new collections. Following relevant accounts and hashtags will keep you in the loop.
- **Discord:** A popular platform for NFT communities, Discord offers spaces where you can engage with like-minded individuals, ask questions, and get advice. Many NFT projects create dedicated servers for their communities, where you can join live chats, participate in events, and learn from experienced NFT creators.
- **Reddit:** Subreddits like r/NFT and r/CryptoArt are great for discussing trends, asking questions, and discovering new projects. Reddit provides a collaborative space where users can share resources and insights.

By joining these communities, you'll be able to stay updated on the latest trends, discover new artists, and engage with others who are passionate about the space. It's also a great way to promote your own work if you're a creator, by tapping into the collective energy and enthusiasm of the NFT ecosystem.

PRESENT REAL-WORLD EXAMPLES

Now you have had a chance to take a look at and understand what NFTs are and how they help to empower creators and consumers and have hopefully had a chance to buy one or join a community, let's get your creative juices flowing and do a whistle-stop review of how NFTs are being used in multiple real-world use cases:

NFTs in gaming: The future of virtual economies

One of the most exciting and rapidly evolving sectors for NFTs is **gaming**. The gaming industry has always had a strong digital component, from virtual currencies like V-bucks in Fortnite to in-game skins and items in franchises like League of Legends. NFTs take this concept a step further by allowing players to own, trade, and even monetize their in-game assets. This not only gives players true ownership over their items but also creates new ways for game developers to generate revenue.

EXAMPLE: AXIE INFINITY – A BLOCKCHAIN-BASED GAME WITH REAL ECONOMIC VALUE

Axie Infinity is a blockchain-based game built on the Ethereum network that uses NFTs to represent in-game creatures called **Axies**. Each Axie is a unique, tradeable NFT that can be bred, trained, and battled in the game. Players can buy, sell, and trade these Axies, with some rare Axies fetching thousands of dollars.

In addition to breeding and battling Axies, players can earn **Smooth Love Potion (SLP)** tokens, which are used for breeding Axies or can be sold on exchanges for real money. This play-to-earn (P2E) model has garnered significant attention, particularly in countries like the Philippines, where some players have been able to generate a full-time income through the game.

Axie Infinity is part of a broader trend in the gaming world, where players can truly own digital assets and have a stake in the game's economy. NFTs in gaming are creating new opportunities for players to monetize their time and skills, fundamentally changing the traditional gaming model.

NFTs for brands: Building a powerful IP

EXAMPLE: PUDGY PENGUINS – FROM OPENSEA TO WALMART

Pudgy Penguins launched in July 2021 as a collection of 8,888 unique, hand-drawn penguin NFTs. The adorable and colorful design quickly captured the attention of the NFT community, distinguishing itself from other projects with its unique charm. By leveraging the growing interest in digital collectibles and the rapid expansion of the NFT space, Pudgy Penguins quickly became a well-known name among NFT enthusiasts.

After a series of turbulent events, the project was acquired in 2022 by Luca Netz, an entrepreneur with a background in branding and business. Luca envisioned taking the Pudgy Penguins brand beyond the NFT ecosystem, creating a larger IP that would appeal to a broader audience.

One of the most successful strategies employed by the Pudgy Penguins team was achieving online virality outside of NFT spaces. This translated into a massive content push, with the team creating GIFs that could be shared in any environment, ensuring the unique art and brand of the penguins appeared everywhere. Forty billion views later, Pudgy Penguins were well on their way to becoming a major brand.

Noticing the strong appeal Pudgy Penguins had with younger audiences through their GIFs, Luca and the team recognized the potential to integrate an NFT brand with physical products. This led to the development of 'phygital' merchandise, including children's plush toys and clothing. Pudgy Penguins set their sights on entering the retail market and after achieving strong online sales through platforms like Amazon, the brand escalated its efforts and began stocking shelves in Walmart stores across the United States.

Pudgy Penguins went on to sell over $10 million worth of goods to buyers who may or may not have even known what an NFT was; an incredible feat of brand-building that would not have been possible without years of community engagement in Web3-native spaces. The transition from digital collectibles to physical goods is a strategy that resonates particularly well with younger consumers, who are often quicker to adopt technological innovations that bridge gaps between digital and physical worlds.

NFTs in the music industry: Revolutionizing copyrights and royalties

The music industry has long been plagued by challenges such as low artist payouts, intermediaries taking large cuts, and issues related to IP rights. NFTs have the potential to address many of these problems by providing musicians with a new way to monetize their work and manage royalties.

> ## EXAMPLE: KINGS OF LEON – A GRAMMY-NOMINATED BAND SELLING NFTS FOR MUSIC
>
> In 2021, the band **Kings of Leon** became the first major band to release an album as an NFT. They sold **NFTs** that gave buyers exclusive access to digital versions of their album, as well as VIP concert tickets and limited-edition artwork. These NFTs were sold on platforms like **YellowHeart**, a marketplace for music NFTs.
>
> By using NFTs, Kings of Leon was able to bypass traditional streaming services like Spotify and Apple Music, which take a significant portion of the revenue generated by streams. The band was able to maintain more control over their music, while their fans gained access to unique experiences and digital content that would not have been available through traditional channels.

BENEFITS OF NFTS IN THE MUSIC INDUSTRY

NFTs offer several key benefits to the music industry:

- **Direct Artist-to-Fan Sales:** Musicians can sell their music and digital content directly to fans, bypassing streaming platforms and record labels.
- **Royalties for Resale:** NFTs allow for continuous royalties on secondary sales, ensuring that artists continue to benefit as their work appreciates in value.
- **Enhanced Fan Engagement:** NFTs can offer fans exclusive access to content, live performances, and other experiences, deepening the connection between artists and their audience.

As the music industry continues to embrace NFTs, we can expect more artists to adopt this technology as a way to regain control over their IP and create deeper connections with their fans.

NFTs in intellectual property and patents

In addition to art, music, and gaming, NFTs have the potential to revolutionize the management and protection of **IP**. Currently, IP rights are often complex and difficult to enforce, especially in the digital world, where assets can be easily replicated. NFTs can help to simplify the management of IP by providing a secure, transparent way to track ownership and provenance.

EXAMPLE: MINTING PATENTS AS NFTS

In traditional systems, patents and IP are registered with a central authority, such as the **United States Patent and Trademark Office (USPTO)**, and can be subject to disputes and lengthy legal processes. However, blockchain technology, combined with NFTs, can streamline the process of registering, buying, and selling patents.

A platform like **IPwe** is already exploring the idea of tokenizing patents as NFTs. This would allow inventors and creators to sell or license their patents directly, providing transparency and a more efficient way to manage IP rights. NFTs would allow for the creation of **smart contracts** that automatically trigger royalty payments when a patent is used, ensuring fair compensation for inventors.

By tokenizing patents and IP, NFTs could bring greater liquidity to markets for intangible assets, allowing creators and innovators to monetize their work more effectively.

NFTs and the future of digital identity

Another important and emerging application of NFTs is in the realm of **digital identity**. As we move further into the digital age, managing and protecting our online identity has become a growing concern. NFTs could provide a solution by giving individuals control over their **digital identity** and personal data.

EXAMPLE: SELF-SOVEREIGN IDENTITY (SSI)

The concept of **SSI** is based on the idea that individuals should have full control over their personal information, rather than relying on centralized authorities like governments, banks, or corporations. NFTs can be used to represent SSIs, ensuring that individuals have a verifiable, secure, and portable form of identity that can be used across platforms without compromising privacy.

Platforms like **Sovrin** are working on blockchain-based identity systems that allow users to control access to their personal data. By using NFTs as digital identity tokens, users can prove their identity without the need for a centralized entity to validate it. This can simplify online authentication processes, reduce the risk of identity theft, and protect personal information from misuse.

As the metaverse continues to grow, the need for secure, privacy-preserving digital identities will become even more critical. NFTs will play a central role in allowing individuals to represent themselves securely in virtual worlds, ensuring that their digital assets and personal data are protected. These digital identities could be used for everything from virtual transactions to social interactions, providing a **trusted, decentralized** identity that crosses platforms and ecosystems.

NFTs in fashion: Redefining ownership and luxury

The fashion industry, known for its high demand for exclusivity and luxury, is increasingly integrating NFTs into its business models. NFTs are offering new ways for designers, brands, and consumers to think about digital and physical fashion. In a world where exclusivity drives value, NFTs are becoming the digital key to owning rare, limited-edition items.

EXAMPLE: DOLCE & GABBANA AND VIRTUAL FASHION

Dolce & Gabbana, one of the most iconic luxury fashion brands, embraced the potential of NFTs with the launch of its **Collezione Genesi** collection in 2021. This collection, a mix of both physical and digital items, was auctioned as NFTs. What made this collection unique is that it wasn't just about creating digital assets; it also came with exclusive physical items. Buyers received real-world luxury fashion pieces, but the NFTs themselves represented the unique ownership of these items, serving as both a **digital certificate of authenticity** and a **token for exclusive access**.

NFTs in fashion not only allow brands to authenticate ownership but also offer a way to engage with the growing **virtual fashion community**.

EXAMPLE: RTFKT STUDIOS – A DIGITAL FASHION REVOLUTION

RTFKT Studios, a leading player in the intersection of fashion, gaming, and NFTs, made waves by creating **virtual sneakers** and other fashion items that exist exclusively as NFTs. In 2021, RTFKT partnered with **Nike** to release **Nike-branded NFT sneakers**. These sneakers are not just digital fashion; they represent a unique form of **digital scarcity**, with each pair being limited edition and verifiably scarce through the blockchain.

As augmented reality (AR) and virtual reality (VR) technologies evolve, NFTs will likely play a significant role in reshaping how we perceive fashion in both the digital and physical worlds.

NFTs in real estate: Virtual property and ownership

Another industry poised for disruption by NFTs is **real estate**, particularly **virtual real estate** in metaverse platforms. While physical real estate has long been viewed as one of the most stable forms of investment, NFTs are bringing this concept into the digital world, allowing users to own virtual land and property in a decentralized, blockchain-based manner.

EXAMPLE: THE SANDBOX – VIRTUAL REAL ESTATE POWERED BY NFTS

The Sandbox is a blockchain-based virtual world where players can buy, sell, and build on virtual land using NFTs. Each plot of land in The Sandbox is an NFT, allowing users to prove ownership and trade land with others. This creates a **real estate market** in a digital world where users can build games, experiences, and even monetize their land.

In 2021, high-profile sales of virtual land on platforms like The Sandbox and **Decentraland** broke records, with virtual plots of land selling for millions of dollars.

Beyond just owning land, users can develop their virtual properties into businesses, museums, galleries, or even digital concert venues.

This has huge implications for how real estate will be viewed in the future, particularly as the metaverse and virtual worlds continue to grow in popularity.

EXAMPLE: DECENTRALAND – DIGITAL REAL ESTATE AS INVESTMENT

Decentraland is another metaverse platform that allows users to buy, sell, and develop virtual land using NFTs. Virtual land in Decentraland can be used for anything from creating digital storefronts to hosting virtual events. The platform hosts auctions where users can purchase parcels of land, and the ownership of each parcel is recorded as an NFT on the Ethereum blockchain.

Real estate in the metaverse is starting to garner the attention of large brands and investors. Companies such as **Atari** and **Gucci** are purchasing virtual land to establish a presence in the metaverse.

Virtual real estate is now seen not just as a place to build, but as an investment vehicle in the emerging digital economy.

CHALLENGES AND RISKS OF NFTS

Despite their many advantages, NFTs come with certain risks and challenges that need to be addressed for mass adoption. Understanding these risks is crucial for anyone looking to invest or create in the NFT space.

1. Environmental concerns
    One of the most frequently discussed challenges of NFTs is their environmental impact. Since some NFTs are minted on energy-intensive blockchains, the process of creating and trading NFTs can consume a significant amount of energy. This has raised concerns about the carbon footprint of NFTs and their sustainability in the long term.

    However, several **Layer 2 solutions** and blockchain platforms, such as **Polygon** and **Tezos**, are working on more energy-efficient methods for minting and trading NFTs. These platforms use alternative consensus mechanisms, such as **Proof of Stake (PoS)**, which significantly reduce energy consumption.

2. Market volatility
    NFT prices can be extremely volatile, with some assets appreciating dramatically in value, while others crash. This speculative nature of NFTs makes them a high-risk investment, and it's important for buyers and creators to understand the risks involved before entering the market.

3. Legal and regulatory issues
    One of the key challenges of the NFT space is the legal and regulatory landscape, particularly around issues of **ownership** and **IP**. While NFTs represent ownership of a unique digital asset, this does not necessarily equate to ownership of the underlying IP rights.

    For instance, if someone purchases an NFT of a digital artwork, they own the **tokenized representation** of that artwork, but they do not automatically own the rights to reproduce, sell, or use the artwork for commercial purposes unless these rights are explicitly granted by the creator. This has led to **legal disputes** over the ownership and use of NFTs, particularly when it comes to **copyright**.

EXAMPLE: CRYPTOPUNKS AND COPYRIGHT ISSUES

One famous example of an NFT legal issue is the case of **CryptoPunks** – one of the first NFT art collections to gain widespread attention. CryptoPunks are 10,000 unique pixelated characters created by software developers **John Watkinson** and **Matt Hall** of **Larva Labs**. While buyers of CryptoPunks NFTs own the tokens representing the art, they do not automatically have copyright ownership of the valuable art itself.

This distinction between token ownership and copyright ownership has raised questions about the resale of NFTs and whether buyers are granted the right to commercially exploit the digital assets they purchase. Some creators have addressed this by explicitly stating that NFT buyers receive certain **usage rights**, while others maintain stricter control over how their work can be used.

As the NFT market continues to expand, **clearer legal frameworks** will be needed to define and protect the rights of creators, buyers, and platforms.

FUTURE OF NFTS: WHAT'S NEXT?

The future of NFTs is full of exciting possibilities as they continue to reshape how we think about ownership, value, and digital identity. While we are still in the early stages of adoption, the potential of NFTs is unfolding rapidly, with new use cases and innovations emerging every day.

As blockchain technology evolves, we can expect NFTs to become more interoperable, seamlessly moving across different blockchain ecosystems. This increased flexibility will allow users to interact with NFTs in a broader range of platforms, ensuring that they retain their value and utility. NFTs will also become more programmable, allowing creators to integrate added functionality such as unlocking exclusive content, granting real-world benefits, and offering dynamic, personalized experiences. This will open up new opportunities in areas like entertainment, virtual experiences, and membership communities.

One of the most exciting developments is the integration of NFTs with decentralized finance. As NFTs become more widely used, they will begin to play a larger role in DeFi protocols, enabling new opportunities for liquidity, lending, and fractional ownership. NFTs could be used as collateral in DeFi platforms, allowing owners to unlock liquidity without losing ownership of the asset. Fractional ownership will democratize access to high-value digital assets like art and virtual real estate, making them more accessible to a wider range of investors.

Looking forward, NFTs are set to become a cornerstone of digital identity; we have already seen the initial use cases. In virtual worlds and the Metaverse, NFTs could represent an individual's identity, linking their online presence to their digital assets. This will allow people to take control of their own digital life, creating new ways to experience ownership and value in a decentralized world.

NFTs will likely become more embedded in mainstream culture, extending beyond art and collectibles to encompass gaming, entertainment, and even social good. With NFTs enabling new ways to engage with creators and brands, they will continue to disrupt industries and open new avenues for interaction, investment, and self-expression. As the technology matures, NFTs will be essential to the way we own, trade, and interact with digital assets, offering a more personalized and equitable digital economy.

In the coming years, it is expected that NFTs will redefine not only how we perceive digital ownership but also how we interact with the broader world of Web3. The future is bright for NFTs, and as we continue to explore the possibilities, they will undoubtedly play an integral role in shaping the decentralized future of the internet.

ACTIONABLE STEPS : HOW TO GET STARTED WITH NFTS

NFTs are rapidly becoming a foundational component of the Web3 ecosystem. Whether you are a creator looking to mint your first NFT or an investor seeking to build a collection, the time to dive into the world of NFTs is now.

Reflection Questions

1. **What does ownership mean to you in the digital world?**

   - Reflect on your current understanding of ownership in the digital realm. What value do you place on owning a digital asset? Consider how NFTs change the notion of ownership compared to physical assets. Do you see NFTs as a way to own something unique, or do you think digital assets will always feel less tangible than physical items?

2. **How might NFTs change the way you engage with your favorite creators or brands?**

   - Think about your interactions with creators, artists, and brands. How would NFTs enable deeper or more meaningful engagement? For example, could owning an NFT give you access to exclusive content, experiences, or special offers? What do you think could be the next step for your favorite brand to integrate NFTs into their business model?

3. **How would you use NFTs to build a community or network?**

- NFTs are not just about owning assets; they can also represent a community. Many NFT collections give holders access to exclusive online communities, events, or governance opportunities. How could you leverage NFTs to create or join a community? What kind of digital experiences would you value being a part of?

4. **What are the risks and rewards of investing in NFTs?**

- NFTs can be volatile, and the market can fluctuate greatly. Reflect on the balance between risk and reward when it comes to investing in NFTs. What level of investment are you comfortable with, and what are you hoping to gain – whether it's a financial return, exclusive content, or new ways to connect with creators?

5. **Explore NFT marketplaces**

- Once you have your wallet set up, start exploring **NFT marketplaces** such as **OpenSea**, **Magic Eden**, **Rarible**, or **SuperRare**. These platforms allow you to buy, sell, and discover a wide range of NFT assets, from digital art to music, virtual real estate, and collectibles. As a collector or investor, use these platforms to discover new NFT drops, view trending collections, and assess the value of different NFTs.

6. **Create and mint your first NFT**

- If you are a creator, minting your first NFT is the next step. Platforms like **OpenSea**, **Rarible**, and **Mintable** make it easy to upload your artwork, music, or digital creation and mint it into an NFT. The process involves uploading your file, filling in metadata (such as title, description, and tags), and paying a small minting fee (usually in ETH) to store your NFT on the blockchain. Once minted, your NFT is officially yours, and you can list it for sale on the marketplace.

7. **Start building your NFT portfolio**

- If you're looking to invest, start by carefully researching NFT projects and creators. Look for collections that resonate with you – whether it's art, music, gaming, or virtual real estate – and evaluate the potential long-term value. Pay attention to the community surrounding each project, as a strong, active community can help drive demand for NFTs. Build your portfolio thoughtfully, and consider starting small to familiarize yourself with the market dynamics.

NFTs are reshaping the way we think about ownership, value, and creativity in the digital world. As blockchain technology continues to evolve, the possibilities for NFTs will expand, offering even more opportunities for creators, collectors, and investors. Whether you're interested in digital art, gaming, virtual real estate, or IP, NFTs offer a revolutionary way to engage with the digital economy.

By following the actionable steps above, you can begin your journey into the world of NFTs and take advantage of the countless opportunities it offers. Remember, this is just the beginning – NFTs are paving the way for a decentralized, more inclusive digital economy that is opening doors for creators and consumers alike.

Chapter 6

DAOs unpacked

Revolutionizing governance and collaboration

In the evolving landscape of Web3, decentralized autonomous organizations (DAOs) are emerging as a transformative force, reshaping the way we think about governance, collaboration, and organizational structures. DAOs introduce a novel model of collective decision-making, powered by blockchain technology that enables communities to self-organize and govern themselves without the need for traditional centralized authorities.

This chapter will provide a thorough understanding of what DAOs are, how they operate, and the key advantages they offer over conventional organizational frameworks. We will also explore real-life use cases and share practical steps you can take to engage with and participate in the growing DAO ecosystem.

WHAT IS A DAO?

A **DAO** is an organization that operates through smart contracts on a blockchain, allowing decisions to be made collectively by its members. Unlike traditional organizations, which are typically governed by a central entity or executive board, DAOs are fully decentralized and rely on a **token-based governance system**. This means that anyone who holds the DAO's native token can participate in the decision-making process, voting on proposals and contributing to the future direction of the organization.

DAOs function without intermediaries, which is one of their most attractive features. Through the use of blockchain technology, DAOs offer a **transparent, trustless,** and **immutable** structure for organizing and executing decisions. All rules governing the DAO – whether they relate to voting, decision-making, or resource allocation – are encoded in **smart contracts**, which are self-executing agreements that automate processes.

DOI: 10.1201/9781003616504-6

WHAT IS A DAO? (IN SIMPLE TERMS)

A DAO is like a club or group where **everyone** gets to have a say in what happens, and no one person or boss has all the control. Imagine you and your friends want to start a club where you make decisions together, like what events to hold or what projects to support. Instead of having one leader, each person gets a **token** (like a special digital ticket), and you use these tokens to vote on decisions.

Everything that happens in the club is **written into a computer program called a smart contract**, which ensures everyone follows the rules. This program is like a set of instructions that run automatically, so the club doesn't need a boss or manager to make things happen.

Since it's all on the **blockchain** (a secure and transparent digital ledger), everyone can see what's happening in the club, and nothing can be changed without the group agreeing. It's a way to organize and make decisions together without needing someone in charge.

THE HISTORY OF DAOS

The concept of DAOs emerged alongside the rise of blockchain technology. The first DAO was created in 2016 on the **Ethereum blockchain**. This initial version of the DAO, known simply as 'The DAO,' aimed to fund projects that would help develop the Ethereum ecosystem. With its **decentralized fundraising model**, The DAO was able to raise an impressive 12.7 million ETH (approximately $150 million at the time), representing 14% of Ethereum's circulating supply.

Unfortunately, a vulnerability in The DAO's smart contract code led to a **hack**, where a portion of the raised funds were stolen. This event sparked a debate within the Ethereum community about the **immutability** of the blockchain and whether it should be altered to reverse the effects of the hack. Ultimately, this controversy led to the **hard fork** of the Ethereum blockchain, creating Ethereum (ETH) and Ethereum Classic (ETC). Despite this setback, the event proved that DAOs were a viable concept, and since then, DAOs have gained significant traction and have been adopted across various industries.

HOW DAOS DIFFER FROM TRADITIONAL ORGANIZATIONS

The most striking difference between DAOs and traditional organizations is the **decentralized nature** of governance. In a traditional organization, decision-making power typically rests with a few top executives or shareholders. This centralized structure can often lead to inefficiencies, lack of accountability, and even corruption, as power is concentrated in the hands of a few.

In contrast, DAOs distribute power among all token holders. Instead of relying on top-down hierarchies, DAOs empower their communities to make decisions collectively, fostering a more **inclusive, equitable**, and **transparent** environment. All decisions made within a DAO are visible on the blockchain, making the process completely transparent.

DAOs also operate with **flexibility** and **fluidity** that traditional companies often lack. In traditional organizations, roles are often rigid, and employees are expected to follow a predefined job description. In DAOs, members can contribute to any project or initiative they are passionate about, based on their skills and values. This flexibility allows for **greater creativity** and **innovation**, as contributors can take on roles that align with their interests and expertise.

BUILDING A DAO: THE CORE COMPONENTS

Creating a DAO is not just about writing code; it's about **building a community** around a shared vision and goal. The first step in launching a DAO is to define its **mission** and **values**. The mission statement serves as the guiding force for the DAO, providing a clear direction for the community to rally behind.

Once the mission is defined, the next step is to **develop a whitepaper**. A whitepaper is a detailed document that outlines the DAO's goals, governance structure, tokenomics, and financial plan. It's essentially the roadmap for the DAO's success. This document helps ensure that all contributors and investors understand the DAO's objectives, how it plans to achieve them, and how the community can participate.

Tokenomics is another crucial component of DAO creation. Tokenomics refers to the economic model of the DAO, including the supply, distribution, and utility of its native token. The token often serves as a **governance tool**, enabling holders to vote on proposals and decisions. It can also be used as an incentive mechanism to reward contributors and ensure the sustainability of the project.

Once the basic framework is in place, the DAO can begin to use **smart contracts** to automate operations. Smart contracts are self-executing agreements that automatically execute actions when certain conditions are met. For example, a smart contract can be used to automatically distribute rewards to contributors or allocate funds to specific initiatives based on community voting.

DAO GOVERNANCE: HOW IT WORKS

One of the defining features of a DAO is its governance system. Unlike traditional organizations, where decisions are made by a central entity or a board of directors, DAO governance is **decentralized** and involves the active participation of its members. The governance of a DAO is usually based on **token-weighted voting**, meaning that the number of tokens a member holds determines their voting power. However, the governance system is flexible, and different voting models can be used, each with its own pros and cons.

1. **One wallet, one vote**
    The **one wallet, one vote** system is one of the most **democratic** voting models in DAOs. In this model, every wallet holder has an equal say in decision-making, regardless of how many tokens they hold. This ensures that every member has a voice, making it fair for all.

    However, the **main challenge** with this model is that it is vulnerable to manipulation. **Sybil attacks** (where a single individual creates multiple wallets) can skew the voting process, giving one person the ability to cast multiple votes. This undermines the fairness of the system, and thus, **while democratic**, it may not always be practical for larger DAOs.

2. **Weighted voting**
    In contrast, **weighted voting** allows for greater influence for those who have contributed more to the DAO. The more tokens a member holds, the greater their voting power. This system rewards **early investors** or those who have made substantial contributions to the DAO's success.

    While this system is more reflective of the **contributions** of each member, it has its drawbacks. The biggest challenge is that it can lead to **centralization**. In large DAOs, members who hold a substantial amount of tokens can dominate the decision-making process, creating an oligarchic structure rather than a democratic one. This can discourage smaller contributors and reduce the overall **equity** of the organization.

3. **Quadratic voting**
    To address the issue of centralization, **quadratic voting** was introduced as an alternative voting model. In quadratic voting, members can allocate more votes to proposals they care about, but the cost of voting increases quadratically as they buy more votes. For example, while the first vote might cost 1 token, the second vote would cost 4 tokens, the third vote 9 tokens, and so on (as illustrated in Figure 6.1).

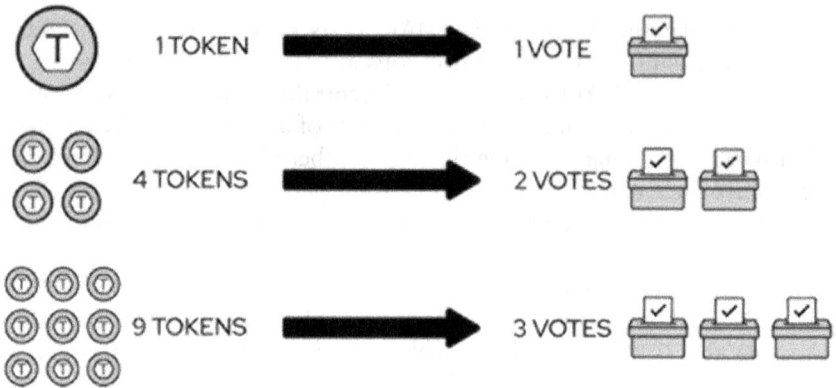

Figure 6.1 Quadratic voting is a special way for a DAO's members to vote on proposals. Instead of each vote costing the same, each extra vote you cast on the same proposal costs more than the last one. This means if you really care about a certain proposal, you can put more votes behind it, but it becomes more expensive each time. It helps balance out the influence in the DAO so that decisions reflect not just what most people want, but also how strongly they feel about it.

This model allows members to **express the strength of their preference** for a particular proposal without giving too much power to large token holders. The cost of acquiring additional votes discourages centralization, and encourages a **more balanced** decision-making process. However, while quadratic voting can be more equitable, it can still be complex for new users to understand and can sometimes lead to **resource imbalances** if some members can afford to buy more votes.

4. Liquid democracy

Liquid democracy takes a more **flexible** approach by allowing members to **delegate** their voting power to others. If a member is not confident in making a decision on a particular topic, they can delegate their voting power to someone they trust, who can vote on their behalf. This creates an **expert-driven** decision-making process for technical or niche subjects, where some members may lack the necessary knowledge.

The advantage of liquid democracy is that it enables **specialized governance**, where decisions can be made by people with the right expertise. However, this system also runs the risk of **centralizing power** if too many members delegate their votes to a small group of trusted individuals. This can lead to a scenario where only a few people hold most of the decision-making power within the DAO, undermining its decentralized intent.

5. Transparency in DAO governance

Regardless of the voting model used, **transparency** is paramount in DAO governance. One of the core principles of DAOs is **openness**, and all proposals, votes, and decisions are recorded on the **blockchain**. This ensures that every member has access to accurate, real-time information about the organization's activities, allowing for a high level of **accountability** and **trust**.

Blockchain's **immutable** ledger guarantees that no one can alter or delete any records, ensuring that all decisions are made in full view of the entire community. This transparency fosters **collaboration** and helps prevent manipulation or corruption, making DAOs more resilient and trustworthy than traditional centralized organizations.

DAO governance is centered on collective decision-making that empowers members to directly shape the future of the organization. By using various voting models such as **one wallet, one vote, weighted voting, quadratic voting**, and **liquid democracy**, DAOs create a governance system that is flexible, democratic, and transparent. The choice of governance model depends on the values of the DAO, balancing factors like equity, specialization, and decentralization. In the end, **transparency** remains a key principle, ensuring that all actions and decisions are accessible to all members, making DAOs unique in their approach to governance and participation.

> *"You need maximal decentralization to build a trust infrastructure."*
> Joseph Lubin (co-founder of Ethereum and founder of ConsenSys)

Of course, in a DAO, voting is an essential part of decision-making, but it's only effective if there is something concrete to vote on. That something is typically a **proposal**. Proposals are the driving force behind any action taken by the DAO, ranging from **new initiatives** and **funding requests** to **protocol upgrades** or any other critical decisions that need to be made. These proposals are submitted by members and become the focal point of the community's decision-making process.

The **proposal process** is at the heart of DAO governance, and its automation through **smart contracts** makes it one of the most efficient and transparent models for decentralized decision-making. When a proposal is made, it is reviewed and discussed by the community, and the voting process is initiated. Once the proposal is approved, the decision is automatically executed by a smart contract, eliminating the need for intermediaries or central authorities.

The **proposal lifecycle** in a DAO typically follows a multi-step process that ensures the proposal is well thought out, thoroughly discussed, and democratically voted upon. This ensures that the entire community has a say in the decisions that shape the future of the DAO. Below are the key stages in the proposal lifecycle:

1. **Drafting**
   The proposal process begins with the **drafting stage**, where a member or group of members submits an idea or plan for a decision. This proposal typically outlines the **specifics** of the decision, including **goals, costs, timeline,** and **expected outcomes.** The member drafting the proposal must provide enough information to help others understand the importance and impact of the decision. It could include, for example, how a funding request will be used, what resources are needed, and how success will be measured.

2. **Discussion**
   After the proposal is drafted, it moves into the **discussion stage.** This is the most dynamic and interactive part of the process. Members of the DAO discuss the proposal in forums or live chat platforms like **Discord,** where they can express their thoughts, ask questions, and raise concerns. The goal of the discussion stage is to ensure that every member has the opportunity to understand the proposal, consider its implications, and share feedback. This open dialogue helps to refine the proposal, address potential issues, and build consensus.

3. **Voting**
   Following the discussion period, the proposal enters the **voting stage.** In this phase, members use the DAO's established governance model to cast their votes on whether the proposal should be approved. The specifics of the voting process, such as **token-weighted voting** or **quadratic voting,** depend on the DAO's governance structure. For example, in token-weighted voting, the more tokens a member holds, the more influence they have on the outcome. In quadratic voting, members can allocate more votes to proposals they care about most, helping to balance the power of larger holders. The voting stage ensures that decisions are made collectively, with the input of the entire community.

4. **Execution**
   Once the proposal is approved by a majority vote, the process moves into the **execution stage.** This is where the power of smart contracts shines. If the proposal involves actions like **funding a new initiative** or **deploying a protocol update,** the smart contract automatically executes the decision. For example, if the proposal was to release funds for a new project, the smart contract would trigger the transfer of funds to the designated recipients. Similarly, if the proposal was about a technical upgrade to the DAO's code, the smart contract would automatically deploy the new code to the blockchain. This

automation ensures that once a decision is made, it is implemented without any delays or manual intervention, making the entire process efficient and transparent.

The proposal lifecycle in a DAO provides a **structured yet flexible framework** for members to submit, discuss, vote on, and execute decisions in a decentralized manner. The transparency and automation afforded by smart contracts create a seamless experience that empowers the community and strengthens the DAO's commitment to collective governance.

As DAOs grow in both size and complexity, **scalability** becomes a critical challenge. Managing an increasing number of proposals and ensuring that members remain actively engaged in voting and decision-making processes can become cumbersome. Without careful planning, large DAOs risk becoming inefficient, with governance processes slowing down as the number of participants and proposals rises. To maintain effective decision-making and active participation, DAOs must ensure their governance models evolve to accommodate the scale of their growth while retaining their **decentralized** and **democratic** principles.

One of the most significant hurdles for scaling DAO governance is managing the **deliberation process**. As more members join the DAO, it becomes increasingly difficult to involve everyone in every decision. To tackle this challenge, DAOs have implemented several innovative solutions to streamline governance while maintaining transparency and inclusivity. These are as follows:

- DAO delegation
  One effective strategy that some DAOs have adopted is **delegation**. In this model, members can delegate their voting power to trusted representatives or smaller committees, allowing these individuals or groups to make decisions on their behalf. This system helps reduce the need for every member to be involved in every proposal, making the decision-making process more efficient. However, delegation still ensures that decisions remain aligned with the broader community's interests, preserving the **democratic nature** of the organization. This structure allows DAOs to scale more efficiently by distributing governance duties while keeping the core principles of collective decision-making intact.
- DAO subDAOs
  Another innovative approach is the creation of **subDAOs**, which are essentially smaller, specialized DAOs within the larger DAO ecosystem. SubDAOs focus on specific areas or functions, such as **marketing**, **development**, or **treasury management**, and are empowered to make decisions related to their respective domains. This decentralization of responsibility reduces the burden on the

main DAO, allowing decisions in niche areas to be made without requiring the entire community to vote on every detail. By allowing subDAOs to handle specialized tasks, the overall DAO can operate more smoothly and efficiently, making it possible to scale the organization while maintaining decentralized governance.

By introducing these layered governance structures, such as **delegation** and **subDAOs,** large DAOs can better handle the complexities of scaling while still prioritizing **decentralization** and **transparency.** These solutions allow DAOs to manage growth without sacrificing the efficiency, inclusivity, and democratic values that make them unique.

TYPES OF DAOS

Now that we understand what DAOs are and how they work governance wise, let's see what use cases they have. DAOs can be classified into different categories, depending on their purpose and objectives. The four main types of DAOs (so far) are:

1. **Investment and grant DAOs**

    These DAOs are reshaping the way we think about funding and supporting innovative projects. These organizations are designed to operate like traditional venture capital (VC) firms, but with a major difference: **decentralization.** Instead of a small group of wealthy investors making all the decisions, members of a DAO collectively pool their resources and use a **token-based voting system** to decide which projects to support. This community-driven approach not only democratizes access to funding but also encourages a more diverse range of voices and ideas to shape the future of innovation.

    - Example: **The LAO** – One of the most profound examples of this model is The LAO, an investment DAO that focuses on funding early-stage blockchain projects. What sets The LAO apart is its **transparency and community-led governance.** Members of The LAO pool their funds together, with each contributing financial resources and gaining governance tokens. These tokens grant them the power to vote on which projects or initiatives to support, ensuring that every participant has an equal say in how the funds are used.

      The LAO operates with an ethos that goes beyond just maximizing returns. Its members are often deeply invested in the **long-term success** and **sustainability** of the projects they support, with a strong focus on fostering innovation in the blockchain and decentralized tech space. By leveraging the collective wisdom and expertise of its diverse community, The LAO not only funds promising projects but also helps guide them toward success by providing advisory support and building a network of trusted connections.

The decentralized nature of The LAO also allows it to be more **agile** and **responsive** compared to traditional VC firms. While a traditional VC might take weeks or months to make an investment decision, The LAO can evaluate opportunities and make decisions within a much shorter time frame, thanks to the efficiency of its decentralized governance structure. This enables The LAO to invest in **emerging technologies** and **disruptive ideas** early on, offering a competitive advantage to both its members and the projects it supports.

In many ways, The LAO is a perfect example of how DAOs are **transforming the landscape of investment** by making it more inclusive, transparent, and collaborative. It demonstrates how blockchain technology can be used to create new financial structures that are driven not by a small group of elites but by a global community of like-minded individuals united by a common vision.

2. Social cause and nonprofit DAOs

These DAOs are an innovative way to tackle the world's most pressing social issues by leveraging blockchain technology and the power of decentralized decision-making. These DAOs prioritize **social good** over profit, focusing on causes like **climate change, education, healthcare.** Instead of relying on traditional nonprofit structures, which often face challenges with funding, transparency, and governance, Social Cause DAOs allow communities to **directly fund and manage** initiatives in a **transparent, secure,** and **democratic** manner.

In these DAOs, members collaborate, pool resources, and vote on which social causes to support or what actions to take. The beauty of this decentralized model is that it allows for greater participation from individuals around the world, enabling those who care deeply about a particular issue to **take ownership** and **drive change** through collective decision-making. The decentralized nature of these organizations ensures that no single entity has control over the funds or decisions, which promotes **fairness, accountability,** and **equity.**

Example: **Gitcoin** – One of the standout examples of a Social Cause DAO is **Gitcoin**, a platform designed to fund open-source projects and social impact initiatives that benefit society at large. Gitcoin operates as a decentralized community of developers, creators, and supporters who come together to address global challenges through technology and innovation. The platform is built on a **token-based reward system**, which incentivizes contributors by offering **tokens** for their work on open-source projects, educational resources, or other social impact initiatives.

Gitcoin's unique approach lies in its ability to bring **direct impact** to causes that would traditionally rely on grants or charitable donations. By utilizing its governance tokens, members of the Gitcoin DAO vote on where to allocate funds, decide on the most deserving projects, and collectively determine the direction the DAO should take. This ensures that the **community** – rather than a few decision-makers – drives the impact, and helps ensure that the most meaningful initiatives are prioritized.

Moreover, Gitcoin's platform encourages **collaboration** between developers and social impact organizations. Through Gitcoin Grants, open-source developers can receive funding for projects that address issues like **climate change**, **public health**, **gender equity**, and **human rights**, creating a system where developers are **rewarded** for working on projects that align with social good. By leveraging blockchain technology, Gitcoin ensures that the funds raised are transparent and trackable, giving donors confidence that their contributions are being used as intended.

What Gitcoin exemplifies is how DAOs can be a **catalyst for change**, transforming the way social causes are funded and managed. The platform fosters a **global community** of individuals who are passionate about using technology for the greater good, showing how decentralized models can have a profound impact on addressing complex, large-scale societal challenges.

3. **Product and business DAOs**

    These DAOs represent a new approach to running businesses, where traditional company structures are replaced with decentralized models. These DAOs are centered on **creating and selling products or services**, but what sets them apart is their governance model. Instead of having a centralized leadership team or board of directors making key decisions, in Product and Business DAOs, **the community of token holders** governs the organization. This means that decisions such as **product development**, **service offerings**, **pricing**, and **business strategy** are made collectively by those who hold the DAO's tokens, with votes determining the direction of the company.

    This decentralized structure allows for a more **transparent, democratic**, and **inclusive** approach to business governance. Token holders have a say in **every key aspect** of the organization, allowing them to directly influence decisions and shape the company's future. Unlike traditional businesses, where a small group of executives often makes unilateral decisions, Product and Business DAOs encourage **community involvement** and **participation**, ensuring that all stakeholders have a voice.

The use of blockchain and smart contracts in these DAOs also ensures **transparency** in decision-making, enabling anyone to track votes, proposals, and outcomes in real-time. Additionally, because these DAOs are decentralized, they are **more resilient** to single points of failure, such as executive decisions that may not align with the wider community's interests.

- Example: **Uniswap** – One of the most prominent examples of a Product and Business DAO is **Uniswap**, a decentralized exchange – which we have already cited as a case in prior chapters – built on the Ethereum blockchain. Uniswap is governed by its community of **UNI token holders**, who have the power to **vote** on important proposals that influence the platform's future. This includes deciding on protocol upgrades, changes to the exchange's fee structure, and which tokens should be listed for trading.

  The beauty of Uniswap's decentralized governance is that it allows for **open, transparent** decision-making. All token holders, from large institutional investors to individual retail traders, can participate in the decision-making process. The votes and proposals are recorded on the blockchain, which means they are **immutable, verifiable**, and easily accessible to the community. This ensures that the decisions made are aligned with the interests of the people who use the platform, rather than being driven by a centralized corporate entity focused solely on profits.

  For example, if the community wants to make an upgrade to the platform – such as integrating new features to improve the user experience or enhancing liquidity pools – they can submit proposals and vote on them. If the proposal receives enough support, it is automatically implemented, often through smart contracts, without the need for intermediaries or centralized approval.

  What sets Uniswap apart is its **open-source, permissionless** nature. Anyone can participate in the ecosystem by becoming a token holder or using the platform, making it a truly **community-driven exchange**. This decentralized model has contributed to Uniswap's success as one of the largest and most popular DEXs in the market, with billions of dollars in daily trading volume.

  Uniswap's DAO governance model represents the future of decentralized business structures. It showcases how **blockchain** technology and **token-based voting** can enable businesses to be run by the community, creating **more inclusive, efficient**, and **responsive** organizations. As the decentralized finance space continues to grow, Uniswap serves as a powerful example of how Product and Business DAOs can transform industries by empowering users and stakeholders to directly influence the success of a platform.

4. **Aficionado or onchain club DAOs**
    These DAOs are unique in that they are formed around **shared hobbies or interests**. They focus on bringing people together through their passion for specific areas such as **collectibles, art, sports**, or **entertainment**. These DAOs operate by pooling funds from their members, which are then used to invest in assets or projects that align with their collective interests. What makes these DAOs stand out is the ability to democratize ownership and access to coveted items or experiences that would traditionally be out of reach for individuals. In these DAOs, members contribute funds, usually through buying **tokens**, and collectively vote on how to manage the investments. This allows people to pool their resources to purchase high-value assets like rare **artworks, sports memorabilia, exclusive collectibles**, or even **real estate**, all while maintaining transparency and equity through blockchain technology. The ability for members to participate in decision-making through token-based voting gives everyone a say in how the assets are acquired, used, or sold.

    The beauty of aficionado or onchain club DAOs is that they transform individual passions into **collective ventures**, giving members the ability to **co-own** and **co-manage** shared assets in a way that was not previously possible in traditional models. It also opens up opportunities for individuals to invest in niche areas, whether it's the arts, sports, or even music, without needing to have large amounts of capital on their own.

    Example: **LinksDAO** – A community of golf enthusiasts who came together with a shared vision: to **own golf courses** and make them accessible to their members. The DAO pools funds from golf lovers around the world and uses the contributions to purchase golf courses. By purchasing and managing golf courses, LinksDAO members are not only investing in a shared passion for golf but are also gaining access to over **400 golf courses worldwide**.

    Members of LinksDAO use **governance tokens** to participate in decision-making processes, such as selecting which golf courses to buy, managing the facilities, and even making improvements. In addition to owning and managing the courses, members enjoy **exclusive access** to play at these prestigious golf courses, creating a community of like-minded enthusiasts who can meet, collaborate, and share in the experience of golf.

LinksDAO is a prime example of how an Aficionado DAO brings a **community-driven approach** to a shared passion, allowing members to collectively fund investments that they wouldn't have been able to achieve individually. The concept of **co-ownership**, powered by blockchain technology and tokenization, has made it easier for people to pool their resources, access exclusive experiences, and build a global community of individuals with a common interest.

This type of DAO is a growing trend, as more hobbyists, collectors, and enthusiasts come together to pool their resources for common causes. Whether it's investing in art, sports teams, music rights, or unique experiences, **Aficionado DAOs** are creating new ways for people to access their dreams and passions in ways that are both financially rewarding and deeply engaging.

In the future, expect to see more DAOs in this space, as people realize the power of **decentralized communities** and the potential for building **shared economies** around personal interests and hobbies. LinksDAO demonstrates how a simple passion for golf can turn into a global network with lasting impact, opening the door for others to follow suit in different niches.

CHALLENGES OF DAOS

While DAO offers revolutionary potential in terms of governance, community-driven decision-making, and decentralization, they are not without their challenges. To realize the full potential of DAOs, several obstacles need to be overcome. These challenges are significant but not insurmountable, and addressing them will be key to the long-term sustainability and growth of the DAO ecosystem.

1. Legal and compliance issues
   DAOs exist in a **legal gray area** in many countries and jurisdictions. Traditional organizations often have a recognized legal structure, such as a corporation or limited liability company (LLC), which allows them to enter into contracts, form partnerships, and manage funds. However, DAOs, due to their decentralized nature, do not fit neatly into existing legal frameworks. Without a legal structure, DAOs face challenges in executing actions such as signing contracts or forming business partnerships.

Furthermore, the lack of a formal structure creates **taxation and compliance uncertainties**. Tax authorities in many countries are still figuring out how to apply existing laws to decentralized organizations. Members of DAOs may face questions about their individual tax liabilities, especially when funds are raised or disbursed through the organization. Some DAOs have tried to navigate these challenges by setting up legal entities in specific jurisdictions, but this may not solve the fundamental issue that DAOs operate outside traditional legal boundaries. For DAOs to truly thrive, there needs to be more clarity and regulation around how they interact with legal systems globally.

2. Security risks

    The reliance on **smart contracts** is both a strength and a vulnerability for DAOs. Smart contracts, which automatically execute predefined actions when certain conditions are met, are central to DAO governance. However, if a smart contract is poorly written or has vulnerabilities, it can be exploited by malicious actors, leading to significant financial losses. One of the most infamous examples of this risk occurred in 2016, as we have already seen earlier in the chapter, when **The DAO** – one of the first major investment DAOs – was hacked due to a vulnerability in its smart contract.

    The DAO hack raised awareness about the need for rigorous **security audits** before deploying smart contracts in any DAO. Even small bugs or overlooked vulnerabilities can lead to catastrophic consequences. Ensuring the security of smart contracts is essential for any DAO, and this requires constant vigilance, expert reviews, and regular updates to the codebase.

3. **Community engagement**

    At the heart of any DAO is its **community** –members who actively participate in decision-making and contribute to the organization's initiatives. However, maintaining high levels of engagement can be challenging. Since DAOs are decentralized, they rely on their members to participate in voting and discussions, ensuring that decisions are made collectively.

    If a DAO's **community becomes disengaged**, it can lead to a decline in participation, making it difficult to form decisions and execute the DAO's objectives. Additionally, some DAOs struggle to make their decision-making process simple enough for members to understand, leading to complexity and frustration. Over time, this lack of engagement and overwhelming complexity can hinder the DAO's ability to function effectively, as members might opt out or disengage from critical decisions. Ensuring that the community remains active, engaged, and motivated to participate is crucial for the success of any DAO. This challenge requires DAOs to find ways to incentivize participation, create clear and transparent processes, and ensure that the voices of all members are heard.

4. Scalability

As DAOs grow and attract more members, **scalability** becomes a major challenge. Smaller DAOs might have a relatively small group of engaged members, making it easier to manage proposals and decision-making. However, as a DAO expands to include hundreds or thousands of participants, the **decision-making process** can become more cumbersome. Managing large-scale voting, resolving disagreements, and ensuring that every proposal is properly discussed can result in delays and inefficiencies.

This challenge is particularly relevant when DAOs rely on voting mechanisms, such as **token-weighted voting** or **quadratic voting**, where each decision might require significant discussion, deliberation, and approval from a large and diverse membership base. Without a robust governance framework that can efficiently scale, the decision-making process could slow down or become chaotic. To solve this, DAOs must develop scalable governance models, such as **delegation** systems, **subDAOs**, or **automated voting tools**, to ensure that growth doesn't lead to inefficiencies or disenfranchisement of members.

THE FUTURE OF DAOS

The future of DAOs is filled with immense potential, as they continue to redefine the landscape of governance, business operations, and social collaboration. While they face challenges in areas such as legal compliance, security, and scalability, ongoing advancements in blockchain technology and governance models offer promising solutions.

As DAOs evolve, we can expect increased mainstream adoption, with more traditional industries embracing decentralized decision-making. The integration of DAOs into various sectors, including finance, nonprofit, and entertainment, will further highlight their transformative capabilities. With enhanced community engagement and improved infrastructure, DAOs have the power to create more equitable, transparent, and democratic systems, offering a glimpse into a decentralized future where individuals are empowered to shape the decisions that affect their lives.

The road ahead for DAOs is unclear and sure to be a bumpy ride with plenty of hurdles to face, but their potential to drive innovation and foster a more inclusive society will make them a cornerstone of Web3 and the future of digital governance – something companies would do well to take seriously.

ACTIONABLE STEPS : HOW TO GET INVOLVED
WITH DAOS

1. Educate yourself about DAO governance

The first step to understanding DAOs is to educate yourself about how they function and the governance models they use. DAOs are decentralized and can have a variety of governance structures, such as **token-weighted voting, liquid democracy**, or **quadratic voting**, each with its own advantages and drawbacks. Platforms like **DAOstack, Aragon**, and **MolochDAO** offer valuable resources for understanding the technical aspects of DAO creation, management, and decision-making. Explore these platforms to get a deeper grasp of how DAOs make collective decisions and what governance models best align with the types of DAOs you want to engage with. Familiarizing yourself with these resources will give you a strong foundation for participating in DAOs more effectively.

2. Join a DAO community

One of the best ways to get started in the DAO space is to **join a DAO community**. Find a DAO that aligns with your interests or values, whether it's focused on art, finance, social causes, or tech. DAO communities are typically active on **Discord, Telegram**, or even **X** (formerly Twitter), where members engage in discussions, share ideas, and collaborate on projects. Once you find a DAO that resonates with you, start by participating in their forums or chat groups.

Ask questions, listen to others, and understand their culture and operations. Platforms like **DAOhaus** and **DeepDAO** offer directories of active DAOs, making it easier to find communities that match your values and goals. Connecting with a DAO's community will help you gain a sense of the DAO's mission and activities before diving into more active participation.

3. Start contributing to a DAO

DAOs are always in need of **contributors** who bring different skills to the table. Whether you're a **developer, marketer, designer**, or have expertise in other areas like finance or project management, there's a place for you in a DAO. Start by reviewing the DAO's website or community forums to find out what types of contributions are needed. Many DAOs have open calls for volunteers or contributors, and it's a great way to get your feet wet. Offering your time, skills, or even resources in exchange for governance tokens will allow you to become a more integral part of the DAO, helping to shape its direction and success. By contributing, you'll not only add value but also gain more experience and influence within the community.

4. Participate in DAO governance

Once you're an active member of a DAO, the next logical step is to get involved in **DAO governance**. This means participating in the voting process by using your **governance tokens**. Every proposal a DAO makes, whether it's about funding, changes to the governance model, or new projects, needs to be voted on. As a token holder, you can read and analyze proposals, participate in discussions, and cast your vote based on what you believe is best for the DAO. Being active in governance is one of the most important ways to make your voice heard and help steer the organization in the right direction. It is the best way to take a hands-on role in shaping the future of the DAO.

5. Consider starting your own DAO

If you have a **decentralized project idea** and feel ready to take a more active role in Web3, consider creating your own DAO. Start by defining the mission, purpose, and goals of your DAO. A well-crafted **whitepaper** will lay out the vision, governance model, tokenomics, and other important details. Platforms like **Aragon** and **DAOstack** provide user-friendly tools and frameworks that allow anyone to launch a DAO with minimal coding expertise. You can design the DAO's voting system, set up funding mechanisms, and ensure that governance is handled efficiently and transparently. Launching a DAO is an exciting way to bring an idea to life in a decentralized environment.

6. Network and collaborate

DAOs are fundamentally built on community and collaboration, so **networking** is essential. Attend DAO-related **meetups, conferences**, and **virtual events** to connect with like-minded individuals. These events are great opportunities to learn from others, explore new ideas, and forge friendships with other DAO members. Engaging with people who share your passion for Web3 technologies can open up new doors for **partnerships, project collaborations**, and **future contributions**. Whether you're attending events hosted by specific DAOs or broader Web3 conferences, networking is a key element in growing within the DAO ecosystem and staying on top of the latest developments.

By following these steps, you can smoothly enter the world of DAOs, making meaningful contributions and building lasting relationships within this **decentralized ecosystem.** Whether you're looking to participate in governance, start a DAO, or simply network with others, the DAO space is full of opportunities to get involved.

Chapter 7

Exploring the Metaverse

Entering new virtual realities

As digital technology continues to evolve at an unprecedented pace, we are standing on the brink of a new era – an era that will redefine how we engage with the internet, changing how we socialize, conduct business, and even shape our personal identities. This emerging reality is known as the **Metaverse** – a vast, interconnected virtual universe that seamlessly blends the physical and digital worlds into one immersive experience. It is a transformative space, powered by cutting-edge technologies such as **blockchain, virtual reality (VR), augmented reality (AR)**, and **NFTs (non-fungible tokens)**, which together enables users to engage with a rich, interactive online environment like never before. It is Web3 at its finest.

The Metaverse represents more than just a collection of virtual spaces or platforms; it is envisioned as a fully immersive digital ecosystem that allows users to not only **connect with others** but also **own digital assets**, participate in shared experiences, and create entirely new realities. Within this ecosystem, individuals will have the opportunity to express themselves, socialize, create, and even conduct business in ways that feel as real as their experiences in the physical world. Whether it's interacting with friends, exploring virtual environments, buying virtual land, or attending immersive events, the Metaverse promises to reshape human interaction and digital experiences in ways we can only begin to imagine.

What makes the Metaverse truly revolutionary is its ability to combine the best of multiple technologies and create a dynamic, user-driven space where individuals are not passive participants but active creators. **Blockchain** ensures that ownership and transactions are transparent, secure, and decentralized, empowering individuals to truly own their virtual assets. **NFTs** provide the unique, digital certificates of ownership for everything from virtual art and collectibles to digital real estate and gaming items, enabling an entirely new economy built on digital ownership. **VR** and **AR** bring immersive, lifelike experiences to users, blurring the lines between physical and virtual realms and enabling them to participate in activities that go beyond the limitations of traditional screens.

DOI: 10.1201/9781003616504-7

This chapter will explore the intricacies of the Metaverse – its defining characteristics, the technologies behind it, and the opportunities it offers across various industries. We'll examine how the Metaverse is already transforming entertainment, gaming, and social media, while also delving into its potential to reshape education, business, and commerce. Through a deeper understanding of how the Metaverse operates and the vast opportunities it presents, you'll gain insight into why this digital revolution has the power to redefine our relationship with technology and the very concept of reality itself.

THE EMERGENCE OF THE METAVERSE

The Metaverse concept was first coined in the 1992 science fiction novel **Snow Crash** by Neal Stephenson, where it depicted a virtual world accessed via the internet. However, it was not until the last decade, with the rise of VR technologies and blockchain-powered assets like **NFTs**, that the Metaverse began to really take shape. In recent years, the term has gained significant traction, largely due to advancements in VR and AR, as well as the growing interest in digital economies and decentralized governance.

At its core, the Metaverse is about creating **immersive, virtual spaces** that transcend the limitations of our physical world. For example, in the Metaverse, you can socialize with others in virtual spaces, attend events, and even own real estate – each of which would be verified and secured on a blockchain.

These spaces aren't just entertainment-focused; they are **interconnected**, meaning users can travel from one virtual environment to another without interruption, all while maintaining ownership of their digital assets. Whether it's buying virtual clothing for your avatar, attending a virtual concert, or meeting someone for a business transaction, everything in the Metaverse is integrated, creating a **holistic digital experience**.

> *"The Metaverse is a collective and persistent digital universe that coexists with and transcends physical reality."*
> Neal Stephenson (author of *Snow Crash*)

CORE TECHNOLOGIES POWERING THE METAVERSE

The Metaverse is not a single platform but a combination of technologies that collectively create immersive, interactive, and decentralized environments. Some of the key technologies driving the development of the Metaverse include the following.

- **Blockchain technology:** At the core of the Metaverse lies blockchain technology, which serves as the foundational layer for decentralization. By utilizing blockchain, the Metaverse enables the secure ownership of digital assets, transparent transactions, and decentralized governance through mechanisms such as decentralized autonomous organizations (DAOs). The blockchain acts as a ledger, ensuring that data such as ownership of virtual land, digital goods, or even interactions between users is tamperproof and accessible to all participants. It also facilitates smart contracts, which are self-executing agreements that automate processes in a way that doesn't require intermediaries, such as banks or online platforms.

  One of the most powerful innovations brought by blockchain to the Metaverse is the NFT. NFTs represent ownership of unique digital assets, which can be anything from virtual real estate to digital art, in-game items, and even collectibles. These assets are providing the foundation for a digital economy where users can truly own, trade, and benefit from their digital possessions.

- Example: **Decentraland** – One of the best examples of a blockchain-powered virtual world in the Metaverse is **Decentraland**. Operating on the **Ethereum blockchain**, Decentraland offers an immersive experience where users can buy, sell, and develop virtual land. Each parcel of land in Decentraland is represented as an **NFT**, meaning that the ownership and transfer of land is recorded and secured on the blockchain. This guarantees that landowners have verifiable and transferable rights to their properties, something that was previously unfeasible in traditional virtual worlds, which were controlled by centralized entities.

  Decentraland operates much like a real-world city – users can build anything on their land, whether it's an art gallery, a nightclub, a marketplace, or even a casino. The possibilities are limitless, and everything created within Decentraland belongs to the creator, ensuring that they can monetize their assets and ideas in new ways. The platform is governed by a **DAO**, where users with governance tokens can vote on important decisions, such as development proposals and platform upgrades, further decentralizing control and making the community an active participant in shaping the platform's future.

  The concept of **digital ownership** is key to Decentraland's success and to the broader Metaverse. Just like owning physical real estate in the real world, owning virtual land in Decentraland provides individuals with valuable resources that can be developed and monetized. Users can rent out their land to businesses or create exclusive experiences that attract visitors and generate revenue. This form of ownership extends to many types of digital assets, not just land – users can also own virtual items like clothing for their avatars, artworks, and collectibles, all of which are secured by NFTs.

The **Decentraland Marketplace** allows users to buy and sell virtual goods with ease. It's a thriving digital economy powered by blockchain technology, where transactions are seamless, transparent, and secure. As Decentraland grows and more users join the platform, the demand for virtual land and digital assets is expected to rise, driving the value of properties and goods up, creating opportunities for users to profit in ways that were not previously possible in online spaces.

Decentraland's integration of NFTs and blockchain technology exemplifies how ownership in the Metaverse can be as legitimate and verifiable as ownership in the physical world. It demonstrates how **blockchain-based economies** are creating new forms of wealth and opportunity, especially in spaces that are decentralized, where control lies not with a central authority but with the community itself. This shift from centralized platforms to decentralized spaces is the key to the Metaverse's potential to transform industries ranging from entertainment, gaming and education.

As the Metaverse continues to evolve, platforms like Decentraland will play a significant role in reshaping how we view digital ownership, virtual economies, and the possibilities for human interaction in online spaces. The lessons learned from Decentraland will pave the way for future advancements in the Metaverse, helping to create a world where digital experiences, interactions, and assets are as valuable and impactful as those in the physical realm.

- **VR:** VR is perhaps the most defining aspect of the Metaverse. With the help of devices like **Oculus Rift, Oculus Quest 2, HTC Vive,** and **PlayStation VR,** users can fully immerse themselves in virtual environments. These devices enable users to interact with and experience the virtual world in 3D space, enhancing the realism and social connectivity that the Metaverse promises. VR platforms also allow for highly interactive experiences, from playing games to attending virtual events.

- Example: **Oculus Quest 2** – This headset is a key player in making VR accessible to a wider audience, allowing users to experience the Metaverse in a way that feels natural and immersive. Unlike many other VR systems, the Quest 2 offers a standalone experience, meaning users don't need to rely on a computer or console to power their virtual environment. This innovation has helped lower the barrier to entry for VR, making it possible for anyone to step into fully immersive virtual worlds with just the headset and controllers.

As VR technology has advanced, so too has the Quest 2's performance. The headset offers stunning resolution and smooth motion, essential for creating a lifelike, believable digital world. The graphics are sharp, and the frame rates are high, reducing any motion sickness and offering a more fluid, interactive experience. This high performance is crucial for a truly immersive Metaverse, where users interact with others in real-time, all within a 3D digital space.

But it's not just the visual and physical aspects of the Quest 2 that contribute to its significance. The headset's integration with platforms like **Facebook Horizon**, which has evolved into **Horizon Worlds**, plays a key role in bridging the gap between VR and social interaction. In Horizon Worlds, users can create digital spaces, attend virtual events, and interact with others as avatars, all within a shared persistent virtual environment. For the first time, VR has become a medium for authentic social engagement, where people can meet friends, collaborate on creative projects, or attend virtual concerts, all from the comfort of their own homes.

The Quest 2 also enables users to explore a variety of virtual experiences that extend beyond gaming, expanding the reach of VR into education, fitness, and digital workspaces. Whether it's stepping into a virtual classroom to attend a lecture, engaging in a fitness class, or working with colleagues in a digital office, the Quest 2 opens up new possibilities for how we interact, learn, and collaborate in the Metaverse. This holistic approach to the virtual world is what makes the Oculus Quest 2 such a transformative device – it's not just about gaming or entertainment, but about reshaping how we connect with each other and the digital landscape.

In summary, the **Oculus Quest 2** is helping to redefine what it means to be part of the Metaverse. By offering a seamless, standalone VR experience, it has made it easier than ever for people to explore virtual environments and participate in a digital economy. As VR technology continues to evolve, devices like the Quest 2 will be central in shaping the future of how we experience and interact with the Metaverse.

- **AR:** While VR creates a fully immersive digital experience, **AR** overlays digital content onto the real world. AR technology enables users to experience virtual elements that are integrated into their physical surroundings, offering a **hybrid experience** that enhances everyday life.

- Example: **Pokémon GO** – **Pokémon GO** is a prime example of how AR has bridged the gap between the digital world and our physical surroundings. Unlike VR, which creates entirely new digital environments, AR enhances our existing world by layering virtual objects into our real-life experiences. Pokémon GO brought this concept to life in a way that captured the world's imagination. In the game, players use their smartphones to find and catch Pokémon that are integrated into their actual environment. Whether it's a park, street corner, or local business these digital creatures appear as if they are right there in the world, thanks to AR technology.

  One of the reasons **Pokémon GO** was so successful is that it encouraged players to engage with both the digital and physical worlds simultaneously. Players didn't just sit in front of a screen – they had to get up, walk around, and explore the world around them to find Pokémon and complete challenges. By turning the entire world into a playground, the game created a unique, real-world social experience, where people of all ages could come together, walk around their neighborhoods, and collaborate or compete in capturing Pokémon. It demonstrated the power of blending physical reality with virtual elements to create a game that felt deeply integrated into everyday life.

  The game also highlighted the potential for **AR to enhance local experiences**. Pokémon GO integrated locations like landmarks, parks, and businesses into the game, encouraging players to visit new places in real life. Many local businesses embraced this by creating 'PokéStops' and offering special in-game rewards to attract players. This blend of gaming, exploration, and commerce demonstrated how AR could bring value not just to users, but also to communities and businesses.

  Beyond gaming, **Pokémon GO** showed the broader implications of AR in the Metaverse. The game was a precursor to how AR could be used for **educational experiences**, **virtual tourism**, or even **virtual retail** – all while being anchored in real-world locations. The success of Pokémon GO proved that AR has the potential to blend the physical and digital, in ways that are both entertaining and practical, paving the way for the future of AR experiences in the Metaverse.

  The lasting impact of **Pokémon GO** is that it revealed the potential of AR to create an interactive, shared experience that brings digital elements to life in the real world.

- NFTs: NFTs are a crucial part of the Metaverse as they allow for the ownership and trading of unique digital assets. As we have seen in the earlier chapter on the topic, NFTs are **blockchain-based** assets that are unique, verifiable, and secure, making them perfect for representing virtual real estate, art, and collectibles within the Metaverse.

The **digital economy**, built upon these unique assets, offers new ways for users to engage in commerce. This economy presents exciting opportunities for both creators and businesses alike.

1. **Virtual real estate**
    Just like the physical world, the Metaverse offers the opportunity to buy, sell, and develop land. Virtual real estate is becoming a hot commodity, as users look to build and monetize properties within virtual worlds. As more users enter the Metaverse, the demand for **virtual land** has grown and should continue to grow exponentially.

- Example: **The Sandbox** – The concept of **virtual real estate** has grown as the Metaverse evolves, becoming a key asset for businesses, creators, and investors alike. Virtual land offers the same opportunities as physical real estate in many ways: it can be developed, sold, leased, and monetized. The ability to create, customize, and control your space in virtual worlds has led to an explosion of interest and investment, especially in platforms like **The Sandbox**.

  The Sandbox is one of the most popular platforms in the Metaverse for buying and developing virtual land. Operating on the Ethereum blockchain, The Sandbox allows users to purchase parcels of land (referred to as 'LAND') that can be used for various purposes, including game development, art installations, virtual events, and immersive experiences. This platform stands out because it not only enables users to buy and sell virtual real estate but also provides powerful tools to develop and interact with that space. Users can build on their LAND using **VoxEdit**, a 3D voxel art editor, and **Game Maker**, a tool for creating interactive games and experiences.

  In The Sandbox, virtual real estate isn't just a passive investment; it becomes a fully customizable and monetizable space. Many entrepreneurs and content creators are taking advantage of this by building virtual stores, art galleries, and entertainment venues. Businesses can host branded experiences, sell virtual goods, or run advertising within these spaces, generating new streams of revenue. One example is **Snoop Dogg**, who created a virtual mansion in The Sandbox where fans can explore, attend concerts, and buy exclusive NFTs. This illustrates how owning virtual land can open doors to unique, scalable business opportunities.

  Moreover, The Sandbox fosters a thriving community where users collaborate, share, and participate in joint ventures. It's not just about owning a piece of land but how you utilize it and create value within the ecosystem. Players, creators, and investors are all part of a dynamic market, which has seen some virtual land plots sell for hundreds of thousands of dollars, underscoring the immense potential of virtual real estate.

As the demand for virtual real estate continues to soar, platforms like **The Sandbox** provide an exciting new frontier for those looking to capitalize on the Metaverse's growth, however we do understand this is an extremely speculative and volatile instrument. Whether you're a business looking to advertise your brand, a creator wanting to bring your vision to life, or an investor seeking the next big asset class, virtual real estate could be a significant consideration. With a vibrant ecosystem and powerful creation tools, The Sandbox is leading the charge in this virtual land boom, and its future promises even more exciting possibilities as the Metaverse continues to develop.

2. Virtual goods and merchandise

Within the Metaverse, virtual goods such as clothing, avatars, and accessories are bought and sold using cryptocurrencies or NFTs. This represents a new way of monetizing digital creativity, with designers and creators able to sell virtual merchandise to users from around the world. Similar to virtual real estate the use of tokens here gives countless opportunities for businesses to create novel revenue streams.

- Example: **Gucci in Roblox** – In the virtual world, brands are finding innovative ways to connect with new generations of digital-native consumers. One of the standout examples of this is **Gucci's collaboration with Roblox**, a popular online gaming platform that allows users to create, share, and explore virtual worlds. Gucci's virtual venture into Roblox highlights the brand's willingness to embrace digital spaces, targeting younger, tech-savvy audiences who spend significant time in the Metaverse.

  Gucci launched an immersive and interactive experience within Roblox called 'Gucci Garden,' a pop-up virtual environment that allowed players to explore Gucci's rich heritage and aesthetic. As users navigated the virtual space, they encountered items designed exclusively for their avatars, such as clothing, accessories, and footwear, all bearing the distinctive Gucci logo. What makes this experience particularly significant is the seamless integration of fashion into the gaming world. It wasn't just about showcasing Gucci's physical designs but offering something entirely digital – a form of virtual self-expression.

Players in Roblox **could purchase these virtual Gucci items using Robux**, the platform's in-game currency, and these virtual goods became valuable status symbols within the game. Gucci's virtual items were not only wearable in the game but served as a new form of social currency – enabling users to show off their taste and exclusivity, similar to how luxury brands operate in the real world.

The Gucci Garden event also offered limited-edition items and exclusive experiences, further emphasizing the rarity and desirability of virtual Gucci products. This scarcity and exclusivity mimic the marketing tactics used in the physical luxury goods market, making the Metaverse an extension of Gucci's high-end, status-driven brand. The success of Gucci in Roblox underscores how brands are using the Metaverse to reach younger audiences in spaces where they are already engaging and spending time, transforming the concept of luxury beyond traditional brick-and-mortar stores into a purely digital and highly interactive world.

By creating a new virtual experience where consumers can purchase exclusive digital items, Gucci has embraced the potential of **virtual goods and merchandise** as a lucrative new revenue stream. It represents a paradigm shift in how businesses, especially in the luxury sector, approach consumer engagement in the digital age. With the rising popularity of the Metaverse, this move into virtual goods is likely to become a significant component of how luxury brands connect with future consumers.

For Gucci, this step not only strengthens its brand in a younger demographic but also sets a precedent for the future of digital fashion in the Metaverse. As the demand for virtual goods continues to rise, other brands are likely to follow Gucci's lead, tapping into the growing market of virtual assets and exploring the boundaries of digital ownership. Gucci in Roblox is a powerful example of how luxury fashion can thrive in the Metaverse, offering a glimpse into the future of the industry, where digital self-expression and status merge seamlessly with the physical world.

3. Virtual event spaces
    When we think about art NFTs they are really interesting case studies for the Metaverse as not only are they digital assets with the ability to be traded, but they lend themselves to the creation of accessible art galleries anyone can attend, irrespective of location.

-   Example: **Decentraland's NFT Art Gallery** – As the Metaverse continues to expand, it is providing exciting new ways to experience art and digital ownership. One of the most innovative ways NFTs are being integrated into the Metaverse is through virtual art galleries. These galleries showcase and sell digital artwork as NFTs, making them a key component of how art is experienced in the Metaverse. A prominent example of this is **Decentraland's NFT Art Gallery**.

The platform's virtual art galleries have become a major hub for digital artists and collectors, offering a unique way for users to experience and purchase NFT art. These galleries are not just simple websites or digital marketplaces; they are immersive, 3D environments where users can walk around, interact with art pieces, and engage with other visitors in real-time. The art displayed in these galleries is often tokenized as NFTs, meaning each piece has a verifiable digital ownership record secured on the blockchain.

In **Decentraland**, artists can showcase their work in virtual spaces, creating dynamic exhibitions that reflect the new era of digital art. Collectors can view the artwork, interact with it, and make purchases directly within the platform. Each NFT serves as a certificate of authenticity and ownership, making it possible for art collectors to own and trade unique digital assets in the same way they would with traditional physical art. This system allows creators to receive compensation for their work while maintaining control over the distribution and ownership of their digital creations.

These NFT art galleries offer several advantages that traditional galleries cannot provide. For one, they are accessible to anyone with an internet connection, making art more globally available. Additionally, the use of **NFTs in Decentraland's art galleries** offers the possibility for royalties to be paid to artists each time their work is resold on the secondary market, a feature that traditional art systems have struggled to implement.

The immersive experience of walking through a **virtual art gallery** in Decentraland brings new dimensions to the concept of art curation. Artists are no longer limited by the walls of physical galleries – they can showcase their work in unique, digital environments that align with the themes of their art. This opens up infinite possibilities for artistic expression and engagement within the Metaverse.

This example of **Decentraland's NFT Art Gallery** illustrates how NFTs are not only creating new markets for digital ownership but are also reshaping the cultural and economic landscapes of the Metaverse. It showcases the future potential of art in virtual worlds and highlights the role NFTs play in facilitating digital ownership, creating new economic models for artists, and enhancing the immersive experience of the Metaverse.

So, now we know the core technologies that are powering the Metaverse and how they are creating new and exciting opportunities, let's cast our net a bit wider and have a look at the future of the Metaverse and its more broad ranging implications in society and for businesses specifically.

THE FUTURE OF THE METAVERSE: WHAT LIES AHEAD?

The potential for the Metaverse is vast, with many industries still in the early stages of exploring its capabilities. As the technology advances, we can expect to see more immersive experiences, increased **virtual commerce**, and greater opportunities for **creators and businesses** to thrive.

- **Expansion of the virtual economy:** The virtual economy will continue to grow, with more businesses creating digital products and services. From virtual fashion to architecture, the possibilities for innovation in the Metaverse are endless.
  - **Opportunities for innovation in virtual markets:** The expansion of the virtual economy will foster a wealth of creative opportunities, empowering businesses to innovate by creating **unique digital experiences** and **assets that could only exist in the Metaverse.**
  - **Integration of virtual and physical markets:** The virtual economy will increasingly intersect with traditional markets, allowing businesses to **bridge the gap between digital and physical goods,** giving consumers access to both virtual and real-world experiences seamlessly.
- Social virtual worlds: Virtual worlds in the Metaverse allow people to interact with one another in shared spaces, offering opportunities for friendship, networking, and even romance.
  - **Metaverse events and experiences:** One of the most popular ways to socialize in the Metaverse is through attending **virtual events.** These could include concerts, live performances, art exhibitions, and even professional conferences.
  - **Metaverse for education:** Education in the Metaverse could be transformative. Virtual classrooms and immersive learning environments allow students to experience complex subjects in new ways, from history lessons where they 'walk' through ancient Rome to interactive STEM labs where they can simulate experiments. The accessibility of such experiences can provide education to people in underserved regions, opening doors to new opportunities and reducing the barriers that often exist in traditional learning systems.

- **Metaverse for health and therapy:** The Metaverse offers unique opportunities for the **mental health** and **wellness** industries. **VR therapy** is already being explored as a way to help patients confront fears (e.g., phobias, PTSD) in controlled virtual environments. It's possible that in the future the Metaverse could host **therapeutic spaces** where users can access counseling or participate in **mental health programs** in a safe and supportive virtual world.
- **Metaverse for charity and social campaigns:** Nonprofits and social enterprises are already exploring the use of the Metaverse to raise awareness and fund social causes. Virtual charity events, auctions, and campaigns can reach a global audience, enabling organizations to make a larger impact.
- **Integration of AI and automation:** As artificial intelligence (AI) continues to evolve, it will likely play a larger role in shaping the Metaverse. AI could be used to create **dynamic, interactive environments** that respond to user behavior, enhancing the realism of virtual experiences.
  - **AI-powered avatars:** In the Metaverse, avatars are digital representations of users, but AI can enhance these avatars' behavior and interactivity. For example, AI can enable avatars to **learn** and **adapt** based on user input, creating personalized experiences. This could range from a virtual assistant avatar that learns your preferences to NPC (non-playable character) avatars that act as intelligent companions in games and virtual spaces.
  - **AI in content creation:** AI can also assist in the **creation of virtual environments.** AI algorithms can generate realistic landscapes, buildings, and even assist in designing avatars or digital art. This would speed up content creation in the Metaverse, allowing creators to focus on higher-level design while automating repetitive tasks.
  - **AI-driven economies:** The virtual economies within the Metaverse can also benefit from AI. From managing virtual supply chains to optimizing virtual marketplaces, AI could power **autonomous economic systems** that predict trends, ensure smooth transactions, and even guide creators on the most profitable virtual goods to develop.
- **The role of DAOs in the Metaverse:** DAOs will likely become central to the governance of virtual worlds. DAOs allow communities to collectively make decisions, manage virtual assets, and govern virtual spaces, ensuring that the Metaverse remains decentralized and community-driven.
  - **Democratizing virtual worlds:** In a Metaverse governed by DAOs, users could vote on how the virtual world operates – whether it's

about implementing new features, setting policies on land ownership, or managing resources. Through DAOs, decisions in the Metaverse could be more **community-driven**, allowing users to have a say in the virtual spaces they inhabit.

- **Creating fairer economies in the Metaverse**: DAOs also enable **fairer distribution of wealth** in virtual spaces. Instead of centralized platforms taking the lion's share of revenue, DAOs could ensure that creators and users are rewarded more equitably, directly receiving a portion of the value they help create.
- **Governance and privacy concerns**: As virtual worlds expand, it's critical that governance structures are **secure**, transparent, and uphold **user privacy**. Blockchain and DAO technology help create systems where all actions are traceable, and rules are predefined, providing more **trust** and accountability in digital governance.
- **The future of work**: With the support of the Metaverse we are likely to see new models for collaboration, allowing employees to work together in virtual offices and meeting rooms, regardless of their physical location. It's a shift toward **remote-first** organizations, but one that takes full advantage of immersive technologies.
  - **Virtual workspaces**: Companies are already experimenting with **virtual offices** where employees can attend meetings, collaborate on projects, and interact with colleagues as if they were in the same room. This shift to virtual offices could drastically reduce overhead costs and increase accessibility for employees around the world.
  - **Metaverse for talent acquisition**: The rise of virtual worlds in the Metaverse could also revolutionize how talent is recruited. **Virtual job fairs** and **immersive recruitment processes** could allow companies to assess candidates in virtual environments, making recruitment more engaging and accessible.
  - **Digital nomadism in the Metaverse**: The Metaverse could further promote **digital nomadism**, allowing workers to set up their offices in virtual worlds, traveling freely while staying productive. The concept of the 'office' will evolve into a **flexible, immersive digital space**, where work can happen anywhere in the world, as long as there's an internet connection.

Of course, despite all these exciting times ahead the path is rarely straightforward, and just as we have seen in the previous chapters of this book related to Web3's emerging technology, there are a range of challenges and hurdles for developers, companies and users to face.

CHALLENGES AND BARRIERS TO THE METAVERSE

Despite its potential, the Metaverse faces several challenges that need to be addressed for it to become a fully integrated and widely adopted ecosystem. These challenges range from technical hurdles to ethical and regulatory concerns.

- **Interoperability:** One of the most pressing challenges in the Metaverse is ensuring interoperability – the seamless ability for assets, avatars, and experiences to function across different virtual worlds and platforms. At present, many virtual environments are siloed, and digital assets such as NFTs are often confined to specific ecosystems. For the Metaverse to achieve its vision of interconnected spaces, overcoming these barriers is crucial. A truly interoperable Metaverse would allow users to transfer assets like virtual clothing or real estate effortlessly across platforms, creating a unified and cohesive experience. Addressing interoperability requires collaboration among developers, companies, and even regulators to establish common standards.
- **Privacy and security:** As digital identities and virtual property ownership become integral components of the Metaverse, the importance of robust privacy and security measures cannot be overstated. With users spending significant amounts of time online investing in virtual assets, the Metaverse presents new risks related to data privacy and cybersecurity. Protecting users' personal information, transactions, and virtual property from malicious actors will be critical in fostering trust and confidence. Furthermore, the decentralized nature of many Metaverse platforms present unique challenges in terms of accountability and security oversight, making it imperative for developers to prioritize strong encryption, transparent systems, and user-controlled privacy settings.
- **Regulation:** As the Metaverse continues to grow and evolve, it is increasingly clear that traditional legal and regulatory frameworks will need to adapt. Issues such as digital property rights, the taxation of virtual assets, and online safety will require new laws and policies tailored to the unique aspects of the Metaverse. Governments face the complex task of ensuring that these frameworks encourage innovation while safeguarding users and ensuring fair practices. Striking the right balance between regulatory oversight and creative freedom will be a key challenge, and policymakers will need to work closely with industry leaders, technologists, and users to create a regulatory environment that supports both growth and protection in this new virtual world.

> *"We are at the beginning of the next chapter for the internet, and it's the next chapter for our company too. The next platform will be even more immersive – an embodied internet where you're in the experience, not just looking at it. We call this the Metaverse, and it will touch every product we build."*
> Mark Zuckerberg (CEO of Meta)

THE BOUNDLESS POTENTIAL OF THE METAVERSE

The Metaverse represents a profound shift in how we experience the digital world, offering boundless opportunities across various sectors – gaming, education, healthcare, business, and beyond. It's not just a virtual playground for digital art and entertainment; it's a vibrant, interconnected ecosystem that has the potential to redefine how we interact, work, learn, and socialize in a fully immersive online space. As technologies like VR, AR, and blockchain continue to evolve, the Metaverse will unlock even greater possibilities for innovation, collaboration, and meaningful engagement.

For both individuals and businesses, the Metaverse offers exciting new pathways to connect and thrive. With no geographical limitations, it fosters more dynamic ways to interact, conduct business, and explore new digital economies. Whether it's creating virtual stores, developing unique experiences, or engaging with global communities, the Metaverse encourages unprecedented levels of creativity and connectivity. It presents a new era where personalized digital interactions bring us closer to one another, bridging gaps and forming new relationships in ways that were once unimaginable.

While the Metaverse is still navigating challenges such as governance, security, privacy, and interoperability, these hurdles are not insurmountable. As decentralized platforms, DAOs, and innovative technologies continue to evolve, the Metaverse will be able to address these concerns, creating an open, secure, and user-driven virtual universe. The continuous development of these technologies will help shape a future where the Metaverse becomes an integral part of the global economy, offering a space where both individuals and businesses can thrive.

Looking ahead, the Metaverse holds immense potential to reshape how we live, work, and play in the digital world. It is a space where creators, entrepreneurs, gamers, and investors can all find opportunities to innovate, collaborate, and succeed. The future of the Metaverse is bright, and as we embark on this exciting journey, it promises to transform how we interact with the digital world and each other. The Metaverse is not just a trend – it is the future, and it is unfolding before us.

ACTIONABLE STEPS: HOW TO GET INVOLVED IN THE METAVERSE

The Metaverse is evolving rapidly, and there are numerous ways to participate, whether you're a creator, developer, investor, or simply an enthusiast
looking to experience what it has to offer. Below are actionable steps you can
take to start your journey into the Metaverse.

1. Understand the Metaverse technologies

Before diving into the Metaverse, it's important to familiarize yourself with the
core technologies that power it. These include **VR**, **AR**, **cryptocurrencies**,
and **NFTs**. Understanding how these technologies work will help you
navigate virtual worlds more effectively and open doors for creativity and
business opportunities.

*   **Action**: Take online courses, attend webinars, or read books/articles
    about VR, AR, cryptocurrencies, and NFTs. Platforms like **Coursera**,
    Udemy, and **edX** offer introductory courses on these technologies.
*   **Example**: Research how the blockchain powers ownership in virtual
    spaces and how VR is used to create immersive worlds.

2. Explore existing Metaverse platforms

Start by exploring existing Metaverse platforms to get a feel for how these
virtual worlds operate. Whether it's socializing, creating virtual assets, or
attending events, these platforms offer a broad range of experiences. Some
popular Metaverse platforms include **Decentraland**, **The Sandbox**,
Roblox, **AltspaceVR**, and **VRChat**.

*   **Action**: Sign up for a platform like **Decentraland** or **Roblox**, create an
    avatar, and start exploring. Attend events, buy virtual assets, or simply
    interact with other users to get a feel for how it works.
*   **Example**: Visit **Decentraland** or **The Sandbox** to view how virtual
    real estate is bought, sold, and developed. This can help you understand
    the growing virtual economy.

3. Invest in virtual assets

The Metaverse economy is powered by virtual assets, which include
land, digital art, clothing, and accessories. These assets are typically
represented by **NFTs** on blockchain platforms, ensuring that ownership is
transparent and secure.

- **Action**: Research NFTs and virtual asset marketplaces like **OpenSea**, **Rarible**, and **SuperRare**. Consider purchasing virtual land or digital art on platforms like **Decentraland** or **The Sandbox**.
- **Example**: Buy virtual land in **Decentraland** and start developing it by creating virtual experiences or hosting events.

4. Get involved in virtual communities

The Metaverse thrives on **community-driven innovation**. To make the most of the Metaverse, it's essential to engage with like-minded individuals, whether you're interested in virtual worlds, digital art, or virtual business. Join forums, Discord servers, and social media groups to meet other users and creators.

- **Action**: Join Discord channels or Reddit communities related to the Metaverse. Participate in discussions, ask questions, and stay up to date with the latest trends and opportunities.
- **Example**: Join the **Decentraland Discord server** or follow **#Metaverse** on X (formerly Twitter) to engage with other users and creators, share ideas, and stay updated on events and developments in the space.

5. Experiment with VR

One of the most immersive ways to experience the Metaverse is through **VR**. VR headsets like the **Oculus Quest 2**, **HTC Vive**, or **PlayStation VR** can help you fully engage in virtual worlds, from socializing and gaming to attending live events or virtual workspaces.

- **Action**: Purchase a VR headset and start exploring VR-based platforms like **VRChat**, **AltspaceVR**, or **Rec Room**. Use it to attend virtual events, socialize, or play immersive games.
- **Example**: Try attending a virtual concert or event in **AltspaceVR** or create a virtual workspace in **Spatial** for collaborative work meetings and team interactions.

6. Explore virtual real estate

The concept of **virtual real estate** is becoming increasingly popular in the Metaverse, with platforms like **Decentraland** and **The Sandbox** allowing users to purchase and develop land. This presents an entirely new way of owning and monetizing virtual property.

- **Action**: Start by researching how virtual real estate works. Explore virtual land marketplaces like **Decentraland's** marketplace or **The Sandbox's** marketplace to buy land, design virtual properties, or even create games and experiences.

- **Example**: In **Decentraland**, you can buy a plot of land, build a digital gallery, and display NFTs for sale. This allows you to interact with users while promoting your virtual art or digital creations.

7. Develop and host virtual events

As the Metaverse grows, more individuals and organizations are hosting virtual events such as conferences, concerts, or social gatherings. Hosting a virtual event allows you to engage with a global audience in a way that physical events simply can't match.

- **Action**: Consider hosting your own virtual event in the Metaverse. Whether it's an art exhibition, product launch, or networking event, virtual spaces like **Decentraland** or **AltspaceVR** offer platforms for creators to engage with attendees.
- **Example**: Host a **virtual gallery opening** for your NFTs in **Decentraland**, or organize a **virtual business conference** using **VRChat** or **Gather**, bringing together experts, speakers, and participants from across the world.

8. Stay informed about the Metaverse's evolution

The Metaverse is still in its infancy, and staying informed about new developments and opportunities will help you stay ahead of the curve. With new platforms, tools, and virtual worlds emerging regularly, it's crucial to keep up with the latest trends.

- **Action**: Follow industry news, subscribe to Metaverse-related newsletters, and attend virtual conferences or meetups to network with thought leaders and other participants in the space.
- **Example**: Subscribe to Web3 newsletters or follow **Metaverse-specific news outlets** to get updates on new projects, partnerships, and platform developments.

The Metaverse represents a new frontier in digital experiences, offering vast opportunities for individuals, businesses, and creators to innovate, interact, and grow. By understanding the key technologies, exploring existing platforms, investing in virtual assets, and engaging with communities, you can actively participate in the development of this exciting new world.

The future of the Metaverse is still being written, and by taking the steps outlined in this chapter, you can become part of its evolution, helping to shape a decentralized, immersive, and innovative VR. Whether you're a creator, entrepreneur, or just curious about the possibilities, there's never been a better time to get involved.

Chapter 8

Building Web3

Strategy and culture

The world of **Web3** is rapidly reshaping business landscapes, creating opportunities, and also creating challenges for organizations. In this chapter, we will explore the crucial concepts of **strategy** and **culture**, and how they can be effectively applied in the Web3 world. We will examine how aligning a well-defined **corporate strategy** with an innovative **corporate culture** can help organizations flourish.

For any organization looking to transition to or thrive in the Web3 ecosystem, integrating strategy and culture is key, because the technology alone just won't work. This chapter will explore the foundational aspects of corporate strategy and culture, followed by their application in the Web3 era.

WHAT IS CORPORATE STRATEGY?

At its core, **corporate strategy** is a roadmap that guides an organization toward achieving its long-term goals and objectives. Strategy is essential for setting direction, defining goals, and allocating resources effectively. It's a framework for decision-making that aligns the company's resources and capabilities with its competitive environment.

A solid strategy enables a company to differentiate itself from competitors, identify growth opportunities, and mitigate risks. For businesses entering the Web3 space, a clear strategy is even more critical. It helps ensure that all efforts – whether related to blockchain, non-fungible tokens (NFTs), decentralized finance (DeFi), or metaverse initiatives – are aligned with the organization's broader business goals.

CASE STUDY: INTEL'S WEB3 STRATEGY

Intel, a leading chip manufacturer traditionally known for its semiconductor products, has made significant moves in the Web3 space by expanding its product portfolio and aligning itself with the evolving digital landscape.

DOI: 10.1201/9781003616504-8

This strategy is not just a response to a new technological wave but a forward-thinking approach to harness the transformative potential of Web3 technologies.

Intel's expansion into the Web3 ecosystem can be seen through its venture into the world of cryptocurrency mining and NFTs. Initially, Intel made waves by developing specialized Bitcoin mining chips. These chips were designed to increase the efficiency and power of mining operations, providing the necessary infrastructure for cryptocurrency networks like Bitcoin. By entering the Bitcoin mining sector, Intel signaled its recognition of the growing demand for computational power in decentralized systems, while also tapping into a lucrative, emerging market.

But Intel's Web3 strategy didn't stop at cryptocurrency mining. In line with its commitment to sustainability and innovation, Intel also explored the potential of NFTs. The company began collaborating with artists and tech startups in the NFT space to integrate blockchain technology with its hardware capabilities. Intel's approach to NFTs is a great example of how a traditional tech giant can diversify and adapt its offerings, creating products that bridge the gap between cutting-edge technology and the digital art world. Their partnership in the NFT ecosystem reflects a commitment to fostering new ways of owning, sharing, and monetizing digital content.

What makes Intel's Web3 strategy particularly notable is how it integrates their core technological expertise with the new frontiers of digital business. Rather than sticking solely to their traditional business lines, Intel ventured into uncharted territory, exploring how it could innovate and stay relevant in the rapidly evolving digital economy. By leveraging its established position as a leading hardware provider, Intel's entry into the Web3 space reflects an adaptive strategy that involves taking educated risks – without straying too far from its core competencies in technology.

Moreover, Intel's strategy is closely tied to sustainability. The company has long been committed to reducing the environmental impact of its products. Its foray into mining technologies wasn't solely about business expansion – it was also about developing energy-efficient chips that reduce the power consumption of cryptocurrency mining, addressing one of the biggest criticisms of blockchain technology.

Intel's Web3 strategy exemplifies a smart, proactive approach to the future of technology. It shows that even established corporations, with decades of legacy, can innovate and pivot to stay ahead of emerging trends, making the right moves to ensure long-term growth and relevance. By embracing Web3 technologies, Intel is not only positioning itself as a key player in the future of DeFi but also creating new opportunities for itself in a world where digital ownership and sustainability go hand in hand.

The Web3 space requires a nuanced strategy, one that balances **opportunities** with **risks,** and ensures alignment with the organization's overall mission and vision. When crafting a Web3 strategy, it's essential to ask critical questions like:

- What new opportunities does Web3 present?
- How do these opportunities align with our current business model and goals?
- What resources and partnerships are necessary to succeed in this space?

Effective **strategic planning** for Web3 businesses involves understanding and capitalizing on new technology trends, while also considering regulatory challenges (which we will come onto in the next chapter), security concerns, and market volatility.

WHAT IS CORPORATE CULTURE?

Corporate culture refers to the shared values, beliefs, and practices that define how an organization operates and how employees behave. It's essentially the personality of an organization, influencing everything from decision-making processes to how employees collaborate and engage with customers.

In the context of Web3, culture plays an especially important role. The decentralized and often transparent nature of Web3 technologies demands that organizations be adaptable, open-minded, and capable of fostering a collaborative environment.

Building a corporate culture for Web3 starts with understanding the unique environment this technology fosters – one that embraces decentralization, creativity, and community-driven innovation. When companies enter Web3, they must create a culture that encourages **open dialogue, experimentation,** and **collaboration** across diverse, often geographically dispersed teams.

In Web3 companies, the culture should empower employees to:

- Take risks and experiment without fear of failure.
- Collaborate across boundaries to foster creativity and innovation.
- Engage openly with customers, stakeholders, and the broader community.

THE POWER OF COMBINING CORPORATE STRATEGY AND CORPORATE CULTURE

While strategy defines the 'what' and 'why' of a business, culture defines the 'how.' When these two elements are aligned, they create a powerful synergy that drives innovation, fosters collaboration, and enables organizations to scale effectively.

Combining corporate strategy with culture has numerous benefits:

- **Shared vision and purpose:** When strategy and culture are aligned, employees understand how their work contributes to broader company goals. This shared sense of purpose enhances motivation and engagement, making it easier to achieve success.
- **Innovation and risk-taking:** A culture that supports risk-taking and experimentation allows for the development of disruptive solutions. However, without a clear strategy, these efforts can lack direction. By combining strategy and culture, companies can harness creativity while staying focused on long-term goals.
- **Psychological safety and trust:** Strategy alone cannot foster innovation – culture plays a key role in creating an environment where employees feel safe to take risks. Psychological safety, where employees are encouraged to speak up, share ideas, and make mistakes without fear of repercussions, is crucial for Web3 businesses. This is particularly important in a decentralized Web3 world, where collaboration across teams is essential.
- Example: **The WD-40 company** – WD-40's success lies not only in its strategy but also in its culture. The company uses a combination of strategy and culture to amplify its business outcomes. Garry Ridge, the former CEO, emphasized that **culture multiplies the effectiveness of a strategy.** According to Ridge, a strategy with a culture score of 50 and a culture score of 50 results in 2,500 (50×50) – compared to 500 if the culture score is just 10 (10×50). This illustrates the powerful amplification of results when strategy and culture align.

> *"Culture is the multiplier that drives the strategy. Without the right culture, strategy is just words on paper."*
> Garry Ridge (former CEO and chair of WD-40)

CREATING A WEB3 STRATEGY FOR YOUR BUSINESS

To develop an effective Web3 strategy, organizations must first gain a deep understanding of how Web3 technologies – such as blockchain, DeFi, NFTs, and the metaverse – can contribute to their goals. This is not just about implementing technology for the sake of innovation; it's about aligning these new tools with your organization's long-term objectives to create real value. Here's how to approach it:

1. **Define your target audience: Understanding Web3 users**
   The first step in creating a successful Web3 strategy is to clearly define your target audience. Web3 is still in its early stages, and the users are often highly engaged **early adopters** who are comfortable with new technologies and understand the principles of decentralization. These individuals are driven by values like **privacy, security,** and **transparency,** which makes them different from traditional consumers.

   For example, a Web3-savvy audience is more likely to be invested in the **ownership** of their digital identities, assets, and experiences. Unlike Web2, where users generally trade their data for access, Web3 users expect to have **direct ownership** over their data and digital assets, with the ability to control who sees and uses their information. This represents a profound shift in expectations that businesses need to consider when crafting their approach.

   A successful Web3 strategy requires that your business model not only caters to these values but also **actively engages with** this community. The audience isn't just passively consuming – they are contributors, creators, and curators within ecosystems. Your business must position itself as a collaborator in these decentralized spaces, building trust through transparency and offering customers **more control** over their digital experiences.

2. **Identify new technology opportunities: Leverage Web3 technologies**
   With your target audience in mind, the next step is to explore the various Web3 technologies that can help meet your business goals. **Blockchain** is the core technology powering Web3, but there are many other complementary technologies, such as **DeFi, NFTs,** and **metaverse platforms,** that you can leverage to create value.

   - **Blockchain** offers a decentralized, **tamperproof ledger** that can transform how your business **records and manages data.** This can help build transparency, reduce fraud, and enable seamless cross-border transactions.
   - **DeFi** opens up new financial possibilities, such as lending, borrowing, and trading without intermediaries like banks. You could consider integrating DeFi to offer **financial services directly** to your customers.
   - **NFTs** allow for unique digital assets that can represent ownership of anything from art to virtual goods to access rights. Companies can create **digital collectible items** or **unlock unique experiences** for customers via NFTs.
   - **Metaverse platforms** provide virtual spaces where businesses can create immersive, interactive experiences for their audience, opening up new avenues for **brand engagement** and **product innovation.**

The key to success here is **integration** – finding ways to combine these technologies into a cohesive experience that fits your business model. The opportunities are vast, but aligning them with your business goals and customer needs is the challenge. Whether it's offering new products, enhancing customer engagement, or creating additional revenue streams, understanding how these technologies align with your objectives is critical for a long-term Web3 strategy.

3. **Develop a roadmap: From vision to execution**
   Once you've identified the key technologies and have a solid understanding of your target audience, the next step is to develop a clear roadmap for how you'll integrate these technologies into your business. Your roadmap should outline the **strategic steps, key milestones,** and **timelines** that will guide the development of your Web3 initiatives.

   A comprehensive roadmap should address:

   - **Technology adoption:** What Web3 technologies will you incorporate first? Will you start with blockchain integration, launching NFTs, or exploring the metaverse?
   - **Internal processes and capabilities:** What changes need to happen within your organization to support Web3? This could involve training staff, restructuring departments, or partnering with technology providers.
   - **Partnerships:** Web3 thrives on collaboration. Your roadmap should identify key partners in the Web3 ecosystem, such as **blockchain platforms, metaverse developers,** or **DeFi service providers,** who can help bring your vision to life.
   - **Customer experience:** How will your Web3 initiatives impact your customer journey? Think about how your customers will interact with your products or services in a decentralized context and design the experience accordingly.

   A clear roadmap ensures that everyone in your organization understands the vision, the role they play in the process, and the milestones needed to reach success. It also allows you to **test and learn** in a structured way – implementing new initiatives gradually to gauge customer response and refine your strategy.

4. **Review and adapt: Stay agile in the Web3 landscape**
   Web3 is a rapidly evolving space. The technologies, user behaviors, and market dynamics are shifting fast, which means that **your Web3 strategy must be flexible** and adaptable. Simply implementing Web3 tools and platforms is not enough; you must be willing to review and refine your strategy regularly based on **emerging trends** and **technological developments.**

   For example, the rise of **new blockchain protocols, regulatory changes,** or even shifts in consumer preferences could alter the landscape significantly. Being able to **pivot and adapt** based on these changes is vital for long-term success in the Web3 world.

Agility doesn't mean abandoning your long-term vision; it means being open to experimenting with new ideas and adjusting your approach when necessary. Create mechanisms within your organization to regularly **assess market trends, gather user feedback,** and **measure the impact** of your Web3 initiatives. This ongoing feedback loop will allow you to stay ahead of the curve and continue delivering value to your customers while staying aligned with your business goals.

The key with your strategy is to understand and play with the technology, but know that the more important element is the mindset and the culture you have within your company that will enable you to deliver on your chosen strategy.

If you struggle, and need a little help defining your strategy you could try one of the best strategy tools out there, a SWOT analysis.

SWOT ANALYSIS: A STRATEGIC TOOL FOR GROWTH AND DIRECTION

When it comes to developing a strategy for any business, particularly in the rapidly evolving Web3 space, a SWOT analysis is one of the most powerful and insightful tools you can use. SWOT stands for **Strengths, Weaknesses, Opportunities,** and **Threats.** It is designed to help businesses gain a comprehensive understanding of both the internal and external factors that could influence their success. By conducting a SWOT analysis, you can identify areas where you have a competitive edge, as well as potential challenges that could impact your growth and stability, as illustrated in Figure 8.1.

The first step in a SWOT analysis is to evaluate **internal factors,** which include the resources, capabilities, and processes within your business. For example, think about your organization's structure, culture, the strength of your team, and the systems in place that support your operations. These are aspects you have control over and can leverage to maximize your competitive advantage. On the flip side, consider any internal weaknesses that might hinder your ability to succeed. Perhaps there are inefficiencies in your processes, outdated technology, or resource constraints that need addressing to improve your position in the market.

Next, a SWOT analysis also considers **external factors** – the elements beyond your control that can influence your business. These include market conditions, customer preferences, the competitive landscape, and broader technological trends. Social, cultural, legal, and macroeconomic factors also play a critical role in shaping your external environment. For example, shifts in consumer behavior or new regulatory developments in Web3 can

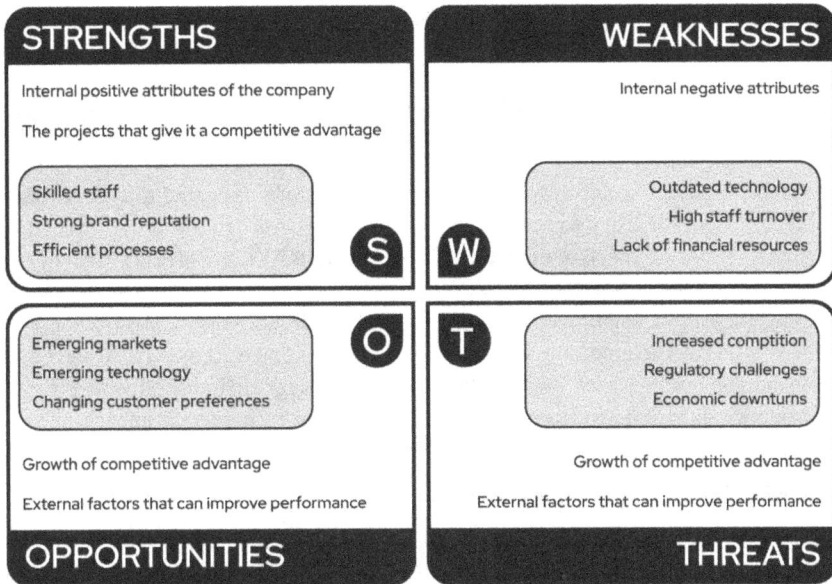

STRENGTHS	WEAKNESSES
Internal positive attributes of the company	Internal negative attributes
The projects that give it a competitive advantage	
Skilled staff Strong brand reputation Efficient processes	Outdated technology High staff turnover Lack of financial resources
S	**W**
O	**T**
Emerging markets Emerging technology Changing customer preferences	Increased comptition Regulatory challenges Economic downturns
Growth of competitive advantage	Growth of competitive advantage
External factors that can improve performance	External factors that can improve performance
OPPORTUNITIES	THREATS

Figure 8.1 A SWOT analysis framework.

create opportunities or pose risks to your business. This section of the analysis helps you spot emerging trends or disruptions that could significantly impact your business model.

In conducting a SWOT analysis, **Strengths** refer to the internal positive attributes of your company that give you a competitive advantage. This might include having a skilled team, a strong reputation, or superior technology. **Weaknesses**, conversely, are the internal challenges or limitations that could undermine your performance – such as high employee turnover, outdated systems, or financial limitations.

The **Opportunities** section focuses on external factors that you can capitalize on to enhance growth and achieve a competitive edge. Think about new market opportunities, technological advancements, or shifts in consumer preferences that you could leverage. **Threats** are the external factors that pose risks to your business, such as economic downturns, regulatory challenges, or intensified competition.

Once you've gathered all of this information, the next step is to **prioritize**. With so many insights coming from your SWOT, it's important to assess what matters most to your business. Which factors have the greatest potential to propel your company forward, and which ones require immediate attention? Should you focus on maximizing the opportunities available in the Web3 space, or would it be more beneficial to address your weaknesses first? Prioritizing your SWOT findings will help you determine the best strategic direction and identify the key areas to focus on for future success.

ADIDAS' WEB3 STRATEGY: PIONEERING VIRTUAL FASHION AND NFTS

Adidas, a global leader in sportswear, has successfully navigated the world of Web3 by strategically partnering with influential players in the decentralized space, including the **Bored Ape Yacht Club (BAYC)**, **PUNKS Comic**, and **Gmoney**. Their innovative Web3 campaign, Into the Metaverse, not only generated millions in revenue but also solidified Adidas' position as a key player in the emerging virtual fashion market. This success was achieved by implementing a clear, well-defined Web3 strategy that aligned with their brand's values and leveraged strategic partnerships to maximize impact.

Defining a clear Web3 strategy

Adidas knew that simply jumping into the Web3 space wasn't enough. The company took a thoughtful approach to ensure that their Web3 strategy was tightly integrated with their core business values and long-term vision. By understanding that **Web3 is not just about digital assets but about new ways of engaging customers** and creating value, Adidas focused on **virtual fashion, NFTs, and decentralized communities** as primary components of their strategy.

They recognized that Web3 technologies – particularly **blockchain, NFTs**, and **metaverse platforms** – could empower them to expand beyond the physical realm and tap into a new generation of tech-savvy, value-driven consumers. Instead of treating Web3 as a separate initiative, Adidas integrated it into their broader marketing strategy, helping them extend their brand narrative into the metaverse while connecting with younger, digitally native audiences.

Strategic partnerships: Tapping into established communities

One of the key reasons Adidas was able to make such a big impact in the Web3 space was their approach to **strategic partnerships**. By aligning themselves with already-established, influential players in the space, Adidas positioned itself as a brand that understood the value of collaboration and community.

- **BAYC**: The collaboration with BAYC, one of the most recognizable and coveted NFT communities, allowed Adidas to reach a broad and highly engaged audience. The partnership helped Adidas gain **visibility in the NFT space** and tap into the social currency that BAYC members enjoy.

Through the partnership, Adidas offered exclusive digital wearables that could be used in virtual environments, combining their reputation in fashion with cutting-edge Web3 technology.

- **PUNKS Comic**: Adidas also joined forces with PUNKS Comic, a digital collectible comic series tied to NFTs. This collaboration introduced Adidas into the world of digital art and storytelling, which is a powerful tool for building brand loyalty in the Web3 space. By aligning with an established NFT project, Adidas made a statement about their commitment to **innovation in digital spaces**.
- **Gmoney**: Adidas also partnered with **Gmoney**, a renowned NFT influencer and collector, further cementing their position in the Web3 space. This partnership brought legitimacy to Adidas' entry into the NFT and virtual fashion markets. Gmoney's influence allowed Adidas to directly connect with Web3 communities that were passionate about **virtual fashion and NFTs**.

NFTs and virtual fashion: Creating unique digital wearables

Adidas' Into the Metaverse campaign offered limited edition **NFTs**, which could be redeemed for digital wearables in the metaverse. These digital assets were not only highly coveted for their exclusivity but also allowed Adidas to showcase their fashion expertise in a completely new context.

The **NFTs** and digital fashion items Adidas offered were designed to be used in virtual environments such as **Decentraland** and **The Sandbox**, two popular metaverse platforms. By creating **exclusive digital items** – like virtual sneakers, hoodies, and other apparel – Adidas allowed its customers to express their style in a virtual space, blending physical products with the growing world of digital fashion.

These digital items could be traded, sold, or worn within metaverse platforms, adding an entirely new revenue stream and creating a **new category of assets** for the company. What Adidas achieved with this is much more than just selling an item; they offered a **unique experience**, where ownership of their NFTs provided both tangible (real-world) and intangible (virtual-world) value.

Revenue generation and market positioning

The Into the Metaverse campaign was not just an experiment in digital fashion – it became a **significant revenue driver** for Adidas. By selling exclusive NFT collections, Adidas not only created buzz and brand awareness but also directly profited from digital asset sales. The limited edition nature of these NFTs and their connection to Adidas' strong brand identity led to high demand and rapid resale, generating millions in revenue within a short period.

More importantly, Adidas was able to establish itself as a **leading brand in the virtual fashion market**, which is expected to grow exponentially as more people engage in digital worlds. This strategy proved that **Web3**, and more specifically **NFTs**, could be highly profitable for businesses willing to embrace the change. The success of Adidas in Web3 is a testament to how traditional brands can thrive by integrating digital innovation with their physical products and services.

Aligning Web3 with Adidas' brand values

From the start, Adidas ensured that their Web3 strategy was aligned with their brand values. Adidas has always been a leader in **sportswear** and **fashion**, and by leveraging Web3 technologies, they were able to tap into the **cultural zeitgeist** surrounding the **metaverse, NFTs**, and **digital ownership**. The company understood that **Web3 is about empowering consumers**, whether it's through ownership of digital assets, participation in decentralized networks, or access to unique, limited-edition content.

In this sense, Adidas wasn't just selling digital items – they were giving their customers an opportunity to **co-create** and be part of something **new and exciting**. This focus on community and empowerment in Web3 helped Adidas to build strong connections with an audience that is increasingly looking for more than just physical products.

Key takeaways from Adidas' Web3 strategy

1. **Strategic partnerships**: By partnering with established Web3 entities like BAYC, PUNKS Comic, and Gmoney, Adidas successfully entered the space and gained instant credibility among Web3 enthusiasts.
2. **Embracing digital fashion and NFTs**: Through the creation of exclusive digital wearables, Adidas was able to blend its traditional fashion expertise with the growing interest in **virtual fashion**, creating an entirely new revenue stream.
3. **Alignment with brand values**: Adidas ensured that their Web3 strategy aligned with their core values, empowering users with a sense of community and ownership while tapping into the metaverse's vast potential.
4. **Revenue generation**: The campaign demonstrated that NFTs and digital goods could be a significant revenue source for traditional brands if implemented effectively.

Now you have your Web3 Strategy, let's take a look at it in action with the following three key pillars – these are perhaps the three most important pillars to consider when building out your roadmap:

1. **Test and learn**
    Typically in business a '**test and learn**' strategy is crucial. We have to take risks and learn from them to get better, though it seems even more critical in Web3 given the pace and scale of disruption. Many companies are entering the Web3 space with low-risk ventures to experiment and understand the landscape before fully committing.
    *   Example: **Intel's Web3 test and learn approach** – As we saw, Intel's approach to Web3 involved testing new technologies gradually, starting with **Bitcoin mining chips** and later expanding into NFTs. Intel's '**test and learn**' strategy allowed the company to gather valuable insights while keeping risks low.
    *   Example: **Starbucks' Web3 test and learn strategy** – Starbucks also adopted a '**test and learn**' strategy when it first ventured into Web3. Their early foray into NFTs was focused on creating limited edition collectible art pieces linked to customer loyalty programs. These NFTs were designed to integrate into Starbucks' loyalty and rewards programs, giving customers the ability to earn special perks and rewards in both the physical and digital worlds.

        The test phase focused on understanding how their customer base would respond to NFTs and whether integrating them into a loyalty program would enhance customer engagement. Starbucks gathered feedback from their community, adapted their offerings, and began expanding their use of NFTs, introducing virtual experiences and collectibles that could be traded or redeemed.

        Starbucks' gradual entry into the Web3 world allowed them to manage the risks associated with new technologies, and it served as a **learning process** that helped them refine their digital offerings. It was a perfect example of how **testing and learning** in Web3 can provide valuable insights without committing large resources upfront.
    This approach allows companies to gain experience and understanding of the Web3 ecosystem, positioning them for success as they scale their operations.
2. **The role of partnerships in Web3 strategy**
    The Web3 ecosystem is fundamentally collaborative. Unlike traditional industries where companies compete fiercely, Web3 thrives

on collaboration, as different projects and platforms are often inter-dependent. Successful Web3 businesses need to engage in strategic collaborations with other companies, platforms, and community members to create innovative solutions and deliver value.

- Example: **Chainlink's partnership with Google Cloud** – **Chainlink,** a decentralized oracle network, entered into a strategic partnership with **Google Cloud** to provide off-chain data to blockchain applications. This collaboration demonstrates how Web3 companies can innovate by leveraging partnerships with established players in the tech industry.
- Example: **Apple and MetaMask partnership** – In a strategic move, **Apple** partnered with **MetaMask,** a leading Web3 wallet provider, to allow users to purchase cryptocurrencies directly from MetaMask using **Apple Pay.** This partnership gave Apple a foothold in the Web3 space, while MetaMask gained credibility from Apple's established brand.
- Example: **Polkadot's cross-chain collaboration** – Polkadot, a multi-chain blockchain platform, enables interoperability between different blockchains. Through cross-chain collaboration, Polkadot fosters innovation by allowing developers to build decentralized applications (dApps) that can operate across multiple blockchain ecosystems.

Collaboration in Web3 is not limited to partnerships with other businesses. It also involves **open-source contributions,** where developers work together to create dApps, protocols, and infrastructure. This collaborative spirit is essential for driving the continuous innovation that characterizes the Web3 space.

3. **Long-term horizons in Web3**
   As organizations transition to Web3, they must adopt a **long-term vision.** Web3 technologies are still in their early stages, and success will not come overnight. Companies must be patient and willing to invest in the future rather than chasing quick returns.
   - Example: **Pepsi's long-term Web3 approach** – Pepsi's **Mic Drop NFT** project was launched without immediate revenue expectations. Instead, Pepsi focused on building its brand within the Web3 ecosystem and engaging with communities. This long-term, strategic approach has positioned Pepsi as a leader in the space, with a more sustainable and impactful presence.
   - Example: **Microsoft's Web3 strategy** – Microsoft has taken a long-term approach to Web3, integrating it gradually into their broader technology ecosystem. One example of this is Microsoft's investment in **decentralized identity** and **blockchain-powered**

cloud services. Rather than rushing into speculative projects, Microsoft has invested in **Azure Blockchain Services**, which enable businesses to integrate blockchain solutions into their operations securely and efficiently.

Rather than focusing solely on quick profit, Microsoft views Web3 as part of a larger **shift toward decentralization**. By providing enterprise-level solutions that enable businesses to build on Web3 principles, Microsoft is positioning itself as a trusted **provider of Web3 infrastructure** over the long term. This strategy allows Microsoft to **continue innovating** and **meeting the needs of businesses** as they transition to decentralized systems while also building a sustainable and future-proof position in the Web3 space.

Now, as we have defined a clear strategy it is also time to define a fitting culture. Neglecting strategy and culture can result in a company losing its way. The lack of a clear strategy can leave companies directionless and unable to adapt to the rapidly changing landscape. Similarly, a poor organizational culture can lead to disengaged employees, stifling innovation and impeding growth.

- Example: **Kodak's failure to embrace digital transformation** – Kodak's reluctance to embrace digital photography and its failure to innovate in the face of technological advancements is a cautionary tale of neglecting strategy and culture. Kodak's corporate culture resisted change, and without a forward-thinking strategy, the company faltered as the market shifted.

So, how do we define our culture?

Start with the three magic questions
To define your culture for Web3, it's important to take a thoughtful approach, considering where you are now, where you want to be, and how you're going to get there. The **three magic questions** can help guide this process, creating a culture that is not only meaningful but also aligns with the transformative power of Web3.

1. **What does it feel like currently to work in your company? What is the culture?**
   Reflecting on the current culture of your organization is a key first step. Does your team feel empowered, trusted, and valued? In Web3, where **decentralization and autonomy are at the heart**, it's important to ask if your current culture supports these principles. Do your employees have the freedom to innovate, share ideas openly, and

collaborate across boundaries, or are there barriers that stifle creativity and initiative? Understanding the feeling of the present allows you to identify the strengths that can be nurtured and the areas that may need to be reimagined to better align with the open, community-driven spirit of Web3.

2. **What would you like it to feel like to work in your company, in the future?**

   In the future, your Web3-driven culture should feel like a place where everyone has a voice, where **transparency** isn't just a value – it's a way of working. You want your team to feel a deep sense of **ownership** over their work, knowing that their contributions matter in ways that go beyond just day-to-day tasks. In a decentralized world, **collaboration** is key, and that sense of trust and freedom is invaluable. Think of a culture where individuals can **experiment, take risks, and explore new ideas without fear of failure,** knowing that their creativity will be celebrated. It's a culture that thrives on **inclusivity,** where everyone can participate and has access to the tools and information they need to make decisions and move forward.

3. **What actions do you need to take in order to make it feel like you want it to? What are the barriers, hurdles, and challenges you need to overcome?**

   To make this ideal culture a reality, you'll need to focus on **empowering your people.** This might mean embracing decentralized leadership, where decisions aren't made by a single authority but by a network of contributors. The tools you use, from digital platforms to communication methods, will play a big role in enabling transparency and collaboration. There may be some challenges in shifting to this way of working, especially if the organization has been more traditional in its structure. It's about creating a supportive environment where everyone feels comfortable embracing new ways of working. Overcoming resistance to change is part of the journey, and it can be done with **clear communication** about why these changes matter and how they'll benefit everyone in the long run. The more you involve your team in the process and allow them to co-create this shift, the more ownership they will feel in the culture itself.

By reflecting on these three questions, you can move toward a culture that is open, decentralized, and full of opportunity for growth and connection. It's not just about adopting new tools or technologies; it's about creating a culture that feels human, empowering, and collaborative.

> *"A company's culture is the foundation for future innovation. An entrepreneur's job is to build the foundation."*
> Brian Chesky (co-founder of Airbnb)

STRATEGY AND CULTURE FOR WEB3: THE ROAD AHEAD

As we close this chapter on strategy and culture in the Web3 space, it's clear that the landscape for businesses is changing rapidly. The Web3 revolution is about much more than just new technologies; it's about embracing a new way of thinking, working, and interacting with the world. To succeed, organizations need to weave together a strong strategy with a supportive and adaptable culture that aligns with the principles of Web3 – decentralization, openness, and community-driven innovation.

A well-defined strategy sets the direction and focus of your Web3 journey. It guides you in exploring opportunities in blockchain, DeFi, NFTs, and metaverse technologies while staying grounded in your core business goals. However, it's the culture that will truly bring your strategy to life. The Web3 space demands more than just technical solutions; it requires a mindset shift toward inclusivity, collaboration, and risk-taking.

Building a culture for Web3 requires a delicate balance. It's about creating a work environment where employees are encouraged to take risks, experiment, and innovate without fear of failure. In this environment, everyone's voice matters, and everyone is empowered to contribute to the company's success. **Transparency, trust, and shared values** form the foundation for this culture, enabling employees to thrive in a decentralized world. But remember, culture doesn't happen overnight; it's an ongoing journey that requires intentionality and effort.

Your strategy should evolve as the Web3 landscape continues to shift, and so should your culture. Adaptability is key. As is the ability to **test and learn,** embracing **long-term horizons,** and staying open to new ideas as they emerge such as partnering with the best technology teams. Whether you're looking to integrate blockchain into your operations, tap into the NFT market, or explore virtual spaces in the metaverse, it's crucial that you approach Web3 with both a solid strategy and a culture that supports innovation.

As you move forward in this exciting new era, think about the **three magic questions** for your culture. By reflecting on your current culture, envisioning your desired future, and identifying the actions needed to bridge the gap, you'll create a culture that's not only aligned with your Web3 strategy but also adaptable enough to succeed in a world of constant change. Together, strategy, and culture are the keys to navigating the Web3 landscape and positioning your organization as a forward-thinking, innovative leader.

The road ahead is full of possibilities. **The future is decentralized, open, and collaborative,** and with the right strategy and culture, your business can thrive within it.

ACTIONABLE STEPS : HOW TO ALIGN STRATEGY AND CULTURE FOR WEB3 SUCCESS

Now that we've explored the importance of **strategy** and **culture** in Web3, and learnt how to build them, it's time to turn theory into practice. Below are actionable steps to help organizations align their strategy and culture to build a successful Web3 business:

1. Start with a clear Web3 vision and strategy

Before diving into the technical details of Web3, it's critical to start with a clear vision and strategy.

- **Action**: Define your organization's long-term vision for Web3. Identify how Web3 technologies – such as **blockchain, DeFi, NFTs,** and the **Metaverse** – can contribute to achieving that vision.
- **Example**: If you are in the entertainment industry, your strategy might focus on utilizing **NFTs** to create unique fan experiences or **blockchain** to build a more transparent and equitable distribution model for content creators.

Key considerations:

- What new opportunities do Web3 technologies offer your business?
- How can Web3 technologies help differentiate your brand and improve your customer offerings?
- What resources and partnerships do you need to bring your Web3 strategy to life?

2. Build a culture of collaboration and transparency

Web3 thrives on **decentralization** and **open collaboration**. For your company to succeed, you must foster a culture that encourages collaboration across teams, communities, and even industries. The culture should be inclusive, transparent, and flexible enough to adapt to rapid changes.

- **Action**: Build a culture that supports **psychological safety** and **open communication**. Employees should feel encouraged to take risks, share ideas, and contribute to the Web3 vision without fear of failure.

Create a space for **peer-to-peer collaboration** and **community-driven innovation**.

- **Example**: Organize regular **cross-functional meetings** where teams from different departments (product, marketing, engineering) come together to brainstorm and discuss Web3 developments. Encourage employees to experiment with new ideas or develop proof-of-concept projects related to Web3.

Key considerations:

- How do you encourage a decentralized approach to decision-making while maintaining alignment with the company's overall strategy?
- How do you ensure transparency in decision-making and business operations?

3. Incorporate decentralized governance into your business model

The Web3 ecosystem thrives on **decentralized governance**. Embrace this principle in your business by allowing **stakeholders** (including employees, customers, and partners) to have a say in the direction of the company.

- **Action**: Consider implementing a **governance token** or **community voting** mechanisms that give your team or community a voice in key decisions. Implement a decentralized autonomous organization-like governance model where decisions are made collaboratively by those who hold a stake in the platform's success.
- **Example**: If you run a Web3 product or service, set up a **community-led advisory board** to help make decisions about product updates, platform features, and governance issues.

Key considerations:

- How will you incentivize stakeholders to participate in governance (e.g., through tokens, voting rights, etc.)?
- What tools or platforms will you use to manage decentralized decision-making and ensure fair participation?

4. Integrate innovation and risk-taking into your culture

Web3 is inherently experimental, and businesses must create a culture that encourages innovation and **calculated risk-taking**. In Web3, failure is often viewed as a **stepping stone to success**, with lessons learned from each iteration.

- **Action**: Create a **culture of experimentation** where your team feels comfortable testing new ideas, trying out blockchain-based models, and exploring Web3 technologies. Develop a clear framework for **evaluating risk** while also encouraging innovation.

- **Example**: Offer **hackathons** or **innovation days** where teams can prototype new Web3 solutions, such as creating an **NFT marketplace**, building a **decentralized finance (DeFi)** protocol, or experimenting with **virtual goods** in the **Metaverse**.

Key considerations:

- How can you ensure that the team understands the risks involved in Web3 ventures while maintaining a culture of innovation?
- What incentives can you provide for risk-taking and innovation?

5. Foster community engagement and collaboration

The Web3 space is all about **community-driven development**, where users, stakeholders, and investors play an active role in shaping the future of the project.

- **Action**: Build and maintain an active, engaged community around your Web3 project. Foster **open communication channels** and **collaborative tools** that allow users to contribute to the project's development and growth. Encourage participation through **incentive models** such as token rewards or governance rights.
- **Example**: Launch a **community bounty program** where users are rewarded for contributions such as bug fixes, marketing efforts, or feature suggestions. This drives engagement and allows your community to help shape the direction of the project.

Key considerations:

- How do you keep the community engaged over the long term, especially after initial hype?
- What tools and platforms can you use to foster communication and collaboration with your community?

6. Adapt your business model to Web3's unique opportunities

Web3 provides new ways to monetize digital assets and services, from **tokenizing assets** to creating **NFT-based products**. Your business model must be flexible enough to accommodate these opportunities, ensuring it stays relevant in the rapidly changing Web3 landscape.

- **Action**: Reevaluate your business model and integrate Web3 opportunities like **NFTs**, **DeFi** solutions, or **tokenized assets** into your product or service offerings. Look for ways to **tokenize ownership** and **leverage blockchain** for transparency and efficiency.

- **Example**: A fashion brand could explore **tokenized fashion** by creating virtual clothing items as NFTs, which customers can purchase, trade, or wear in virtual environments like the **Metaverse**. This adds value to your brand and opens up new revenue streams.

Key considerations:

- What new revenue models (e.g., through NFTs, DeFi, or tokenized assets) can you introduce to your business?
- How do you ensure the transition to Web3 does not alienate existing customers or disrupt your core business model?

7. Ensure long-term sustainability and growth

While Web3 technologies open up exciting opportunities, they also present challenges around scalability, security, and regulation. Businesses need to build long-term **sustainability** into their strategies and cultures.

- **Action**: Focus on developing a **sustainable business model** that incorporates scalable solutions, long-term goals, and a commitment to **ethical practices** in Web3. Prioritize **security**, especially when dealing with sensitive data or financial transactions, and ensure that your project can scale effectively as user demand grows.
- **Example**: Develop a **long-term roadmap** that includes regular upgrades, updates, and improvements to your platform. Implement **security best practices**, such as smart contract audits, to ensure the safety of user funds and data.

Key considerations:

- How will you balance short-term experimentation with long-term sustainability?
- What measures will you put in place to ensure your project's growth remains secure and scalable?

The key to success in Web3 lies in **innovation, adaptability,** and the **willingness to build a transparent, inclusive, and decentralized business.** By aligning your company's strategy with its culture, you'll be positioned to have the best chance of this.

Chapter 9

Navigating Web3 compliance

Ethics, regulation, and innovation

Navigating the complex legal, regulatory, and compliance landscape of Web3 is an essential part of building a successful business in the decentralized world. While Web3 offers exciting opportunities, it also comes with a unique set of challenges, particularly when it comes to legal compliance. Areas such as intellectual property (IP), risk management, and the application of securities and financial regulations are all points of friction with regulators. In this chapter, we will explore the key elements of Web3 compliance, from understanding the regulatory frameworks that govern crypto projects to effectively structuring your business to ensure you meet the necessary legal requirements.

In this chapter we will share a crash course – somewhat in the form of a menu of eleven key items, so that you can jump to the section you are most interested in and take account of the key ideas. Although one may argue that this is the most complex chapter in the book you will find it one of the briefest given the landscape is so rapidly evolving and that readers will be from a range of jurisdictions and thus the content can never be tailored enough nor detailed enough. As such, it is recommended to seek specific legal advice as you encounter any or all of these items in your day-to-day.

1. **Understanding legal, regulatory, and compliance frameworks**
    As of 2025, the legal ambiguity surrounding Web3 is gradually reducing as more jurisdictions introduce explicit regulations. Global regulators are tightening oversight on decentralized finance (DeFi), decentralized autonomous organization (DAOs), and token issuance. Compliance is now a critical factor for project sustainability, with jurisdictions like the **United States, European Union (Markets in Crypto-Assets (MiCA)), United Kingdom (FSMA 2023), Singapore, and South Africa** enforcing clearer rules. Web3 businesses must proactively align their operations with these frameworks as they are necessary for protecting consumers, ensuring transparency, and preventing illicit activities such as money laundering or fraud.

DOI: 10.1201/9781003616504-9

- **Laws:** The laws governing Web3 projects stem from traditional legal systems but adapt to the complexities of decentralized technologies. Different countries follow different approaches, with **common law** systems (e.g., the United Kingdom, United States, Australia) relying on judicial precedents and **civil law** systems (e.g., Germany, France) depending more heavily on codified statutes.
- **Regulation:** Regulations are rules developed by regulatory authorities to ensure that laws are effectively implemented and adhered to. In Web3, this includes **financial services regulations, consumer protection laws,** and **anti-money laundering (AML)** regulations.
- **Compliance:** Compliance refers to the practices and procedures that businesses put in place to adhere to these laws and regulations. It includes **training, governance frameworks,** and **systems and controls** to ensure all activities align with legal standards.

"We count very heavily on principles of behavior rather than loads of rules."
Warren Buffett (chairperson of Berkshire Hathaway)

---

2. **Structuring a Web3 project or entity**
   One of the first steps in launching a Web3 project is deciding how to structure your business. There are several options available, each with its own legal and regulatory implications.
   - **Centralized vs. decentralized structure:** A traditional company structure, such as a **limited liability company (LLC)** or **corporation,** is common in the Web3 space. These entities offer protection from personal liability for business owners and are subject to the laws of the jurisdiction in which they are incorporated. For example, many Web3 projects are structured in places like **Zug, Switzerland,** or **Delaware, United States,** due to their favorable regulatory environments.

   On the other hand, some projects opt for decentralized structures, which may be governed by **DAOs.** In a DAO, decision-making is distributed across the community, and there are no traditional shareholders or employees. This structure aligns well with the ideals of decentralization and can be used to avoid certain jurisdictional regulations. However, decentralized structures may face difficulties with legal contracts, liability, and regulatory scrutiny.

   While decentralized structures, such as DAOs, offer innovative governance models, regulators are increasingly scrutinizing their legal status. Several jurisdictions, including the **United States, United**

Kingdom, and Hong Kong, now require certain DAOs to register as legal entities or comply with financial regulations. Wyoming's 2024 DAO-related legal cases have set precedents, requiring DAOs to clarify legal liability. Projects considering a decentralized governance structure must evaluate whether they can meet compliance obligations, including know your customer (KYC)/AML requirements and liability protections.

- Example: **MakerDAO,** a decentralized lending platform, operates under a governance model where **MKR token holders** vote on key decisions, such as protocol upgrades. While MakerDAO avoids the centralized entity model, it still faces regulatory challenges related to compliance and taxation.
- Example: **The Ooki DAO case (2022, U.S.)** led to **legal action against token holders,** demonstrating that regulators **can hold DAO participants accountable.** As such, DAOs should consider a **hybrid compliance approach,** balancing decentralization with legal protections.

Best Practices for DAO Compliance:
- Registering as a **legal entity** (e.g., Wyoming DAO LLC, Swiss Association).
- Implementing **KYC/AML measures** for governance token holders.
- Ensuring **tax compliance** by properly categorizing DAO income.
- Using **DAO legal wrappers** to shield members from liability.

3. **Applying securities laws to Web3 projects**

A common point of confusion in Web3 compliance is understanding whether tokens issued by a project are considered **securities** under existing financial laws. The application of **securities laws** varies across jurisdictions, but the central question remains: do the tokens represent a **financial investment** with the expectation of **profits** based on the efforts of others? This is critical in determining whether the token falls under securities regulations.

- **US SEC's Howey Test:** In the United States, the **Securities and Exchange Commission (SEC)** uses the **Howey Test** to determine whether a token is a security. The test evaluates whether an investment contract exists by assessing:
  1. **An investment of money,**
  2. **In a common enterprise,**
  3. **With an expectation of profits** that are primarily from the **efforts of others.**

- Example: In 2022, the SEC filed a lawsuit against **Ripple Labs,** alleging that its **XRP token** was a security and that the company had violated securities laws by selling unregistered securities. In 2024, the US SEC **escalated its enforcement actions** against crypto firms, targeting projects that fail to register as securities. The **Ripple (XRP) lawsuit outcome** set important legal precedents, influencing how new projects classify tokens.
- In parallel, the **US Stablecoin Regulation Act,** expected in 2025, aims to regulate stablecoin issuers, requiring them to maintain 1:1 fiat reserves. This means projects launching stablecoins in the United States must secure **federal licensing** or risk legal consequences.
- The **MiCA regulation,** fully in force by 2024/2025, sets strict requirements for Web3 businesses operating in the **European Economic Area (EEA).** Under MiCA, companies dealing with crypto assets, stablecoins, and utility tokens must comply with new **licensing requirements, transparency rules, and consumer protection laws.** The regulation explicitly defines **asset-referenced tokens (ARTs) and e-money tokens (EMTs),** subjecting them to stricter financial oversight.
  - Example: **A project launching a stablecoin in Europe must now hold sufficient reserves and register with a regulated entity to operate legally.** Failure to comply can lead to enforcement actions, including fines and operational bans.
  - MiCA creates a clear structure for the issuance and trading of crypto assets, ensuring consumer protection and market integrity. It also introduces requirements for transparency and disclosure from issuers and operators of **virtual assets.**
4. **E-money and payment services regulations**
   The **Electronic Money (e-money)** regulation is another crucial legal area for Web3 projects to understand. If your token is classified as e-money, it could require regulatory approval. E-money refers to electronically stored monetary value that can be used to make payments or purchases.
   - **E-money regulation:** A token is considered e-money if it represents a claim on the issuer that can be redeemed at par value and used for payments. This is typically relevant for stablecoins, which are often pegged to a fiat currency like the **US dollar.**
     - Example: **Tether (USDT)** is one of the most widely used stablecoins, pegged to the US dollar. Because Tether is used for financial transactions, it may fall under e-money regulations in jurisdictions like the European Union or United Kingdom.

- **Payment services:** If a Web3 business facilitates the transfer or exchange of crypto for fiat or other digital assets, it may also fall under **payment services regulations.** These services are subject to compliance with money transmission and **AML** rules.
  - Example: **Coinbase,** a cryptocurrency exchange, facilitates crypto-to-fiat conversions, which puts it under both **e-money** and **payment services** regulations.
- Governments worldwide are launching **Central Bank Digital Currencies (CBDCs)** – digital versions of fiat currency issued by central banks. CBDCs could **impact Web3 businesses** in the following ways:
  - **Regulatory scrutiny:** Web3 payment solutions must ensure **interoperability with CBDCs** while **complying with stricter monitoring.**
  - **Potential restrictions on decentralized assets:** Some governments may **prioritize CBDC adoption over stablecoins** (e.g., China's crackdown on USDT in favor of the digital yuan).
  - **Privacy concerns:** Unlike decentralized assets, CBDCs **enable full government tracking,** raising concerns for privacy-focused Web3 platforms.
  - Example: **The European Union's Digital Euro and China's Digital Yuan** are being **actively tested.** Web3 businesses must anticipate **CBDC regulations impacting stablecoin usage and DeFi transactions.**

5. Virtual asset service provider laws

   Another key regulatory framework for Web3 businesses is the **Virtual Asset Service Provider (VASP)** laws. These regulations govern firms that offer services related to crypto assets, including exchanges, custodians, and wallet providers. VASPs are required to comply with **AML/counter-terrorist financing (CTF)** obligations.
   - **Global VASP regulations:** Countries like **Hong Kong,** the **Cayman Islands,** and British Virgin Islands **(BVI)** have adopted VASP laws that require crypto firms to register with local regulators and adhere to AML/CTF requirements. These regulations help mitigate the risk of crypto being used for money laundering or other illicit activities.
     - Example: **Binance,** one of the largest cryptocurrency exchanges globally, has faced scrutiny from regulators in various jurisdictions due to its failure to fully comply with VASP regulations. It has been forced to adjust its operations and seek regulatory licenses in regions like **Europe** and **Asia** to continue operating legally.

6. **Intellectual property and data protection in Web3**

   In addition to financial regulations, **IP** and **data protection laws** play a significant role in Web3 compliance. The unique nature of Web3 assets, particularly **NFTs** and **smart contracts**, creates a need for businesses to safeguard their **IP rights** and protect user data.

   - **IP in Web3**: NFTs have revolutionized the way digital art and assets are sold and owned. However, the legal landscape surrounding **NFTs** and their IP implications remains unclear in many jurisdictions. NFTs represent ownership of a digital asset, but the underlying IP (e.g., the digital art) may still belong to the original creator unless otherwise specified in the contract.
     - Example: The **Blockeras** case in 2022, where a company was sued for infringing on Juventus Football Club's trademark in an NFT series, highlights the need for thorough IP clearance when creating and selling NFTs.
   - **Data protection**: Web3 companies must comply with **data protection regulations** like the **GDPR (General Data Protection Regulation)** in the European Union, which governs how user data is collected, stored, and processed. Even though blockchain transactions are decentralized and transparent, they may still involve personal data, particularly when users engage with platforms or services that require identity verification.
     - Example: Projects like **Soulbound tokens**, which are used for identity verification, need to consider the data protection implications of linking personal data to blockchain transactions. Ensuring that data is protected and stored securely is essential to comply with privacy laws.

7. **Compliance and risk management**

   The growing complexity of Web3 regulations has increased the demand for robust **compliance frameworks**. Regulatory authorities expect Web3 businesses to implement comprehensive compliance programs to prevent **money laundering, fraud,** and other financial crimes.

   - **Compliance functions**: Web3 businesses must establish a **compliance function** that educates employees on legal and regulatory requirements and monitors adherence to these rules. The compliance team must act as a second line of defense, ensuring that operational teams are aware of and following the correct procedures.
   - **Risk management**: Web3 projects are exposed to various risks, including **financial, operational,** and **regulatory** risks. Effective **risk management** involves identifying, assessing, and mitigating these risks to protect the business and its stakeholders.

- Example: **Binance** faced regulatory issues due to its insufficient risk management and compliance measures. The company has since taken steps to implement more stringent compliance programs to align with the global regulatory environment.

To meet compliance obligations while preserving Web3's decentralized nature, businesses can leverage **on-chain compliance tools** such as:

- **Chainalysis and TRM Labs:** Provide real-time transaction monitoring for AML compliance.
- **Zero-knowledge KYC solutions (e.g., Polygon ID, zkPass):** Enable identity verification **without revealing personal data,** balancing **privacy with regulation.**

Decentralized compliance frameworks are emerging, offering **automated smart contract compliance** to ensure platforms **adhere to jurisdictional regulations while maintaining decentralization.**

8. Navigating international compliance challenges

Since Web3 is a global phenomenon, businesses in the space often operate across borders. This creates complexities when it comes to **international compliance** – different countries have varying laws regarding crypto, token issuance, decentralized platforms, and consumer protection. To ensure long-term success and avoid legal conflicts, Web3 projects must adopt a proactive approach to managing international regulations.

- **Global regulatory divergence:** While some jurisdictions, such as **Switzerland** and **Singapore,** have implemented **crypto-friendly regulations,** others have banned or severely restricted crypto activities. Understanding these differences is crucial for any Web3 company that wants to operate across multiple regions.
    - Example: **Binance** faced significant challenges in maintaining global operations as different countries imposed restrictions on its services. For instance, the **United Kingdom's Financial Conduct Authority (FCA)** imposed restrictions on Binance's operations, while the company faced similar scrutiny in **Japan, Germany,** and **Thailand.** In response, Binance shifted its focus to regions with more favorable regulations, such as **Dubai** while working to enhance their compliance requirements globally.
- Asia presents a **diverse regulatory environment for Web3 projects. China continues to ban crypto trading and mining,** but **Hong Kong has embraced a regulated crypto framework, allowing retail trading under strict compliance rules.**
    - Example: **Japan's Financial Services Agency (FSA)** has enforced **stablecoin legislation,** requiring issuers to maintain

1:1 **reserve backing** and operate under regulated financial institutions. Monetary Authority of Singapore (**MAS**) is tightening AML/KYC requirements, particularly for **DeFi platforms and NFT markets.**

- Meanwhile, **India's harsh tax policies,** including a **30% capital gains tax and 1% TDS on crypto transactions,** have pushed many projects to relocate elsewhere. Businesses must carefully navigate these regional regulatory differences to ensure legal compliance.

- Africa's Web3 ecosystem is evolving under **new compliance frameworks. Nigeria's SEC** has introduced crypto licensing guidelines, while **South Africa's Financial Sector Conduct Authority (FSCA)** now classifies crypto as a **financial product,** requiring licensing for exchanges. In contrast, **Kenya and Ghana** have tightened scrutiny over crypto firms due to concerns about fraud and Ponzi schemes.

   - Example: The **Central African Republic (CAR)** made headlines by **adopting Bitcoin as legal tender,** though its implementation has faced **regulatory challenges.** Entrepreneurs entering Africa's Web3 market must closely monitor these **diverse regulatory landscapes** to ensure compliance.

- **Compliance in the global DeFi landscape:** DeFi platforms, due to their decentralized nature, often face regulatory challenges in terms of jurisdiction. Many DeFi platforms are not based in any single country, and their decentralized structure means there's no clear 'central authority' that can be held accountable.

   - Example: **Aave** and **Uniswap,** two leading DeFi platforms, have faced scrutiny in certain countries where regulators are concerned about the risks associated with unregulated financial services. These platforms are working on creating solutions that comply with local regulations, while still maintaining the decentralized ethos.

9. **Web3 privacy laws and user rights**

   Web3 businesses must also understand **privacy laws** and ensure that users' data is properly protected, even in decentralized environments. Blockchain's transparency and immutability present unique challenges for privacy. While transactions are publicly recorded on the blockchain, personal data may still be required for compliance or user verification purposes.

   - **GDPR and blockchain:** One of the major regulatory frameworks to consider is **GDPR,** which governs how personal data is handled within the European Union. Web3 projects that process personal data must ensure compliance with these regulations, particularly with respect to **data minimization, right to access,** and **right to be forgotten** – concepts that can be difficult to implement on immutable blockchain systems.

- Example: **Brave Browser,** a privacy-centric browser that integrates with Web3, allows users to interact with ads and websites while protecting their privacy. Brave offers a clear example of how blockchain technology can be combined with **privacy protection.** By limiting the amount of personal data tracked and ensuring user anonymity, Brave complies with privacy regulations while using blockchain technology.
- **Privacy and anonymity in DeFi:** Privacy and anonymity are often seen as essential elements in Web3 projects, especially in DeFi. However, maintaining anonymity while complying with global **KYC** and **AML regulations** is a fine line. DeFi platforms must balance privacy with the need to comply with financial regulations, especially when it comes to **transaction transparency** and **auditing.**
  - Example: **Monero** and **Zcash,** privacy-focused cryptocurrencies, have faced regulatory scrutiny for enabling anonymous transactions, which makes it harder to track illicit activities like money laundering. While privacy is a key selling point for these cryptocurrencies, it also places them in conflict with regulatory requirements designed to combat financial crimes.

10. **Anti-money laundering and counter-terrorist financing regulations in Web3**

    AML and CTF regulations are a cornerstone of the financial regulatory framework for Web3 businesses. They ensure that businesses don't inadvertently facilitate illegal activities such as money laundering or terrorist financing.

    Several jurisdictions, including **the European Union, United States, and Japan,** have either **banned** or **restricted** privacy-focused cryptocurrencies like **Monero (XMR) and Zcash (ZEC)** due to concerns over illicit finance. In 2024, the **European Banking Authority (EBA) recommended the delisting of untraceable digital assets** from compliant exchanges, prompting centralized exchanges to remove them.

    Similarly, global regulators have intensified scrutiny on **non-compliant crypto exchanges,** leading to legal actions against major centralized exchanges. Exchanges failing to implement **robust KYC/ AML protocols** are now at risk of being blocked from operating in key jurisdictions.

    - **AML compliance:** Web3 companies that deal with crypto exchanges, token transfers, or custodial services must ensure compliance with **AML regulations.** This includes verifying user identities (KYC procedures), monitoring transactions for suspicious activities, and reporting certain transactions to authorities.

- Example: **Coinbase,** one of the largest crypto exchanges, has integrated robust **KYC** and **AML procedures** into its platform. This includes identity verification for users and monitoring of transactions to ensure that all activities comply with US and international AML regulations. Coinbase's efforts to maintain regulatory compliance in multiple jurisdictions demonstrates the importance of adapting to local AML/CTF rules.

- **CTF and decentralized governance:** With decentralized platforms and DAOs, enforcing AML and CTF measures can be more challenging. These organizations often lack a centralized authority that can monitor and verify transactions. As DAOs and decentralized projects become more common, there is an increasing need for **smart contract-based compliance** solutions, which can automate and enforce AML/CTF requirements in a decentralized environment.

  - Example: The **FATF (Financial Action Task Force)** has issued guidelines on how DeFi platforms and DAOs should comply with AML regulations. As DeFi platforms become more mainstream, they will need to incorporate compliance measures, such as the use of decentralized identity verification solutions or tools that can automatically flag suspicious transactions.

11. **The role of smart contracts in legal compliance**

    While smart contracts ensure automation and transparency, they also raise concerns about **legal enforceability and liability.** If a smart contract malfunctions or executes incorrectly due to a bug, **who is legally responsible?** Some jurisdictions, such as **Switzerland and Singapore, recognize smart contracts as legally binding,** while others still lack formal legal recognition.

    Additionally, **traditional dispute resolution mechanisms** (e.g., courts, arbitration) are not always applicable to smart contracts. To mitigate these risks, projects can implement **built-in dispute resolution mechanisms,** such as:

    - **Multi-signature execution** (requiring multiple approvals before major smart contract actions).
    - **On-chain arbitration platforms** (e.g., Kleros) for dispute resolution.
    - **Code audits and formal verification** to prevent exploitative vulnerabilities.
      - Example: The 'The DAO Hack' (2016) exposed major flaws in smart contract governance, leading to the Ethereum hard fork. Modern projects should consider implementing **fail-safes** to prevent irreversible consequences.

- **Smart contracts and compliance automation:** One of the primary advantages of smart contracts is the ability to automate compliance with regulations. By programming compliance rules directly into the contract, Web3 businesses can ensure that certain conditions, such as **tax calculations, transaction limits,** or **jurisdiction-specific regulations,** are automatically met without human intervention.
  - Example: **Chainlink**'s decentralized oracle network allows smart contracts to access external data feeds, ensuring that compliance rules and regulatory requirements are automatically adhered to in a transparent and verifiable way.

ENSURING COMPLIANCE IN THE WEB3 ERA

As we continue to explore and embrace the immense potential of Web3, it's crucial to acknowledge the complex and evolving landscape of legal and regulatory challenges that come with it. The decentralized nature of Web3 technologies disrupts traditional industries, presenting both exciting opportunities and unique risks. Businesses venturing into the Web3 space must navigate an intricate maze of legal frameworks, compliance measures, and financial regulations to build a solid foundation for long-term success.

This chapter has provided a high-level overview of the key elements of Web3 compliance. From understanding the foundational legal structures and navigating securities laws to implementing AML and CTF regulations, we've touched upon the vital steps needed to ensure your Web3 project operates within the boundaries of the law. Additionally, we've explored the challenges presented by decentralized systems and smart contracts, such as their enforceability and the need for rigorous compliance with local and international regulations.

While this chapter serves as an introduction to the complex compliance issues Web3 organizations face, it's important to remember that the landscape is dynamic, with laws and regulations continually evolving. Therefore, the insights provided here should not be considered as legal advice but rather as a starting point for your journey into the world of Web3 compliance. It is highly recommended that you seek specific legal advice tailored to your jurisdiction and project to ensure you meet all necessary regulatory requirements and avoid potential legal pitfalls.

In a Web3 ecosystem that is constantly shifting, ensuring compliance is not only about avoiding legal complications. It is about cultivating trust and fostering positive relationships with users, investors, and regulatory bodies. Businesses that prioritize legal adherence stay proactive in responding to regulatory changes and implement strong governance frameworks making them well positioned to thrive in this new digital economy. By approaching compliance with diligence and foresight, you can mitigate risks, unlock new opportunities, and set your project up for sustainable growth.

Web3 represents a remarkable convergence of technological innovation and the need for robust governance. As we move forward, businesses that master the complexities of compliance in this space will not only safeguard their operations but also establish transparency and trust – critical elements that underpin the success of any decentralized project. The journey to building a compliant Web3 business is challenging but essential. By approaching compliance with care, you ensure that your Web3 venture can flourish within the boundaries of the law and continue to thrive in a decentralized future.

The future of Web3 compliance will be shaped by a balance between innovation and regulation. Businesses must be proactive in adapting to evolving laws while maintaining the core principles of decentralization.

LOOKING AHEAD TO 2030

- Expect tighter integration of **AI and compliance automation** in smart contracts.
- Regulatory bodies will standardize global Web3 laws, reducing jurisdictional inconsistencies.
- Governments may push for greater DeFi oversight and regulated DAO frameworks.

By staying ahead of regulatory changes and embracing compliance best practices, Web3 projects can position themselves for long-term success in the decentralized economy.

ACTIONABLE STEPS : ENSURING WEB3 COMPLIANCE

Navigating the compliance landscape in Web3 can be daunting, but with the right strategies, businesses can ensure that they meet regulatory requirements, protect users, and create a trustworthy environment. Here are actionable steps to help you stay compliant in the Web3 world:

1. Understand the legal and regulatory frameworks

Start by developing a comprehensive understanding of the regulatory environment within your operating jurisdiction. Web3 is a global phenomenon, so it's critical to consider both local and international regulations.

- **Action**: Research and stay updated on the evolving legal and regulatory frameworks governing **crypto assets, tokens**, and **smart contracts** in the jurisdictions where you plan to operate.
- **Example**: Ensure compliance with **MiCA** in Europe or **the SEC's guidance** on tokens in the United States. Identify whether your project could be subject to **securities laws, e-money regulations**, or other local compliance requirements.

2. Choose the right business structure

The legal structure of your business will affect how you are regulated. Web3 projects can operate as traditional entities like LLCs or corporations, or take the form of DAOs, each with different compliance needs.

- **Action**: Evaluate your project's needs to determine whether a centralized or decentralized structure is more appropriate. Consider consulting with a legal expert to choose a structure that minimizes legal risks while aligning with your goals.
- **Example**: If you choose a centralized structure, make sure to set up your entity in a jurisdiction with favorable Web3 laws, such as **Delaware (United States)** or **Switzerland**. If you're creating a DAO, ensure that you understand the legal implications of decentralized governance.

3. Comply with securities laws

Understanding whether your token is classified as a **security** is crucial to determining your regulatory obligations. Different countries apply varying criteria for what constitutes a security.

- **Action**: Apply the **Howey Test** (United States) or similar tests in your jurisdiction to assess whether your token constitutes a security. If it is, ensure that you follow the necessary registration and reporting requirements.
- **Example: Ripple Labs** is currently in a legal battle with the SEC, which argues that **XRP** is a security. Understanding these issues can help you avoid costly legal battles and ensure that your token is compliant with financial regulations.

4. Ensure AML/CTF compliance

Web3 projects must adopt robust **AML** and **CTF** procedures to prevent illegal activities on their platforms. This may include conducting **KYC** checks and monitoring suspicious activities.

- **Action**: Implement a KYC/AML process for users engaging with your platform. Use **identity verification tools** and regularly monitor transactions to ensure compliance with international AML standards.
- **Example: Coinbase** has implemented extensive **KYC** procedures, requiring users to verify their identities before they can trade. This helps ensure the platform complies with US and international AML laws.

5. Secure IP rights

Web3 technologies, particularly **NFTs**, are heavily reliant on IP. When creating and selling NFTs or other digital assets, it's crucial to clarify ownership and ensure that creators retain control over their IP.

- **Action**: If you're creating NFTs or other digital assets, ensure that the terms of sale specify what IP rights are being transferred and what remains with the creator. Consider using **smart contracts** to enforce IP rights.
- **Example**: When creating NFTs, specify in the contract whether the buyer is purchasing the NFT with full ownership of the underlying IP (e.g., digital art) or simply purchasing the NFT as a **collector's item** without the right to reproduce or sell the underlying work.

6. Adopt privacy and data protection best practices

Privacy concerns are central to the Web3 space, where decentralized technologies operate alongside traditional privacy laws. You must ensure that any personal data processed by your platform complies with global data protection laws, such as the **GDPR** in Europe.

- **Action**: Implement data protection policies that comply with **GDPR** or equivalent regulations in your jurisdiction. Ensure that any personally identifiable information (PII) is handled and stored securely.
- **Example**: If you're running a platform that collects personal data for KYC purposes, ensure that the data is encrypted and that users have control over their data. This could involve implementing a **self-sovereign identity (SSI)** model, where users maintain control over their identity data.

7. Stay updated on regulatory changes

Regulatory landscapes are fluid, particularly in emerging technologies like Web3. Laws and regulations are continuously evolving, and businesses must stay agile to adapt to these changes.

- **Action**: Create a system for regularly reviewing and staying updated on Web3-related regulations. This could include subscribing to industry newsletters, attending Web3 conferences, and engaging with legal advisors who specialize in the space.

- **Example**: Following **FATF** guidelines and keeping abreast of updates from global regulatory bodies will ensure you're prepared for new developments in the legal landscape.

8. Implement robust risk management practices

Web3 projects face numerous risks, including technological vulnerabilities, legal challenges, and market volatility. Effective risk management will help mitigate these issues and protect your business.

- **Action**: Develop a risk management strategy that includes identifying potential risks, implementing mitigation strategies, and regularly reviewing these strategies as the business and regulatory environment evolve.
- **Example**: **Aave** mitigates risk by using risk management parameters in its DeFi protocol, adjusting borrowing limits and collateral ratios based on market conditions. Similarly, ensure that your Web3 platform has **built-in security protocols** to prevent hacks or unauthorized access.

9. Consult legal advisors and compliance experts

Given the complexity of Web3 compliance, it's essential to seek expert legal guidance to ensure that your project aligns with all relevant laws and regulations. This is especially important as Web3 is still a developing space, with legal precedents continuing to emerge.

- **Action**: Regularly consult with **legal advisors** who specialize in Web3, blockchain, and crypto law to ensure that your business remains compliant. It's also advisable to work with **compliance professionals** to set up internal compliance procedures.
- **Example**: If you're unsure about the regulatory classification of your token, consult a blockchain lawyer who can guide you through the legal intricacies and assist in navigating complex jurisdictions.

By following these actionable steps, all businesses can ensure they are not only legally compliant but also build a foundation of trust and security with their users, partners, and regulators. The rapidly evolving Web3 space requires businesses to stay vigilant, flexible, and proactive in addressing compliance challenges.

Chapter 10

AI meets Web3

Merging intelligence with decentralization

On a busy street in Lagos, a bright-eyed entrepreneur named Yemi hustles between meetings, her smartphone in hand. She's excited about her newly launched mobile application that uses artificial intelligence (AI) to connect local farmers with urban grocers. Just a few short years ago, such a project might have demanded prohibitively expensive technology and a large engineering team. Now, with cloud platforms and open-source AI libraries at her disposal, Yemi has managed to create a powerful tool on a modest budget. Farmers who once struggled to sell surplus produce now find steady urban markets; grocers receive fresher stock, and customers enjoy fewer stockouts. In the pilot phase alone, the app has reportedly cut food waste by 20% and boosted revenues for participating farmers.

Yemi's story is not an isolated phenomenon. Rather, it reflects a broader wave of innovation where AI is forging new pathways in efficiency, creativity, and collaboration. From advanced image recognition in healthcare, to real-time language translation, to financial services that automatically detect fraud, AI has become the driving force behind many industries. Across the globe, AI – as another pillar of Web3 – is transforming how we learn, conduct business, shop, communicate, and organize societies. It's also helping to reshape the digital landscape itself – from the early days of a static Web1 to the interactive world of Web2, and now to the decentralized ethos of Web3.

In this chapter, we explore the story of AI: how it began as a collection of high-reaching ideas, how it weathered setbacks and winters, and how breakthroughs in machine learning propelled it into the mainstream. We will delve into the far-reaching impacts on society, including the rise of new economies, shifts in labor dynamics, and changing perceptions of privacy and ethics. We'll also examine how AI is intertwining with blockchain-based technologies in the emerging Web3 era, paying special attention to the world of decentralized finance (DeFi). As DeFi exemplifies a major use case for blockchain – offering lending, borrowing, and trading services without traditional intermediaries – the potential synergy with AI reveals

DOI: 10.1201/9781003616504-10

how these two revolutionary forces might converge to enhance transparency, autonomy, and innovation.

> *"AI is going to be everywhere. All our applications will be intelligent, all devices will have some intelligence embedded in them – AI will permeate everything we do."*
> Satya Nadella (CEO of Microsoft)

FROM SPECULATIVE CONCEPTS TO MAINSTREAM REALITY

The history of AI is one of big dreams, crushing disillusionments, and then surprising revivals. In the 1950s, pioneers like Alan Turing asked foundational questions: can machines think? That question led to the first wave of AI research, and for a time, optimism soared. Early systems managed small successes, such as basic problem-solving and even game-playing software. Yet as time went on, researchers over-promised capabilities. By the 1970s and 1980s, limited hardware and a shortage of realistic applications led to extended 'AI winters,' in which investment and enthusiasm cooled significantly.

A revival began in the 1990s as technology improved and scientists shifted from symbolic AI (explicitly programmed logic) to data-driven approaches known as machine learning. One notable triumph was IBM's Deep Blue defeating chess grandmaster Garry Kasparov in 1997. Though Deep Blue's specialization to chess was narrow, it showed that AI could excel within a defined domain. In the early 2000s, interest in AI steadily grew, supported by faster processors, greater memory, and a rising tide of digital data. Then in 2012, a dramatic milestone came when deep neural networks – trained on massive image datasets – outperformed competitors in the ImageNet challenge, dramatically reducing error rates in image classification. This success, led by a team at the University of Toronto, signaled the full-blown rise of 'deep learning' and reignited global excitement around AI.

Open-source libraries like TensorFlow and PyTorch emerged soon after, making it easier for developers to experiment with advanced AI algorithms. Cloud services from providers such as Amazon, Alphabet, and Microsoft offered scalable computing power, once reserved only for major research institutions. By the early 2020s, many businesses – ranging from small startups to multinational corporations – began integrating AI for tasks such as fraud detection, supply chain optimization, recommendation systems, and content moderation. A 2023 McKinsey study showed that over half of the world's large enterprises now deploy AI in at least one department.

Leaders began to take notice. Some, like Tesla and SpaceX CEO Elon Musk, voiced caution about potential existential threats. Others, like Marc

Benioff of Salesforce, drew attention to the competitive advantages gained by companies that harness large volumes of data. Responsible or not, the growth of AI continued at a breakneck pace, soon permeating nearly every sector of the economy.

> *"I think we should be very careful about AI. If I had to guess what our biggest existential threat is, it's probably that. But if used responsibly, AI will unlock possibilities we can't even imagine."*
> Elon Musk (CEO of Tesla, X and SpaceX)

SOCIETY AND ECONOMY IN AN AI WORLD

Today, AI has a transformative footprint in multiple areas, from how we manage finances to how we communicate, heal, learn, or even entertain ourselves. The economic potential is massive. Typical management consulting forecasts suggest that by 2030, AI could add up to $15.7 trillion to global GDP. Such numbers stem from multiple pathways: AI can automate monotonous tasks, optimize resource use, enhance decision-making with predictive analytics, and create entirely new product categories.

Within manufacturing, computer vision-based quality checks can decrease defects and reduce production time. Predictive maintenance, fed by sensor data, anticipates machine breakdowns before they happen, saving hundreds of thousands of dollars in unplanned downtime. In finance, high-frequency trading driven by AI executes data-driven decisions at speeds human traders cannot match. AI-based fraud detection sifts through billions of data points daily to spot suspicious activity, adding crucial layers of security to digital banking and crypto transactions.

Healthcare stands out as a key beneficiary. AI-assisted diagnostics help radiologists spot tumors earlier in mammograms or CT scans, boosting survival rates. Personalized treatment applications continuously learn from patient feedback, adjusting medication or therapy to better manage chronic diseases like diabetes or hypertension. Beyond these clear advantages, AI's reach extends to realms like design, music, agriculture, energy management, and scientific research.

Yet, each stride in efficiency and innovation also has consequences. AI's ability to perform tasks once handled by people sparks debate on labor, wages, and the overall future of work. However, just like with the rise of the internet, most estimates indicate that the number of jobs destroyed will be far less than the arising new jobs – at least, for the foreseeable future. Roles such as data scientists, machine learning engineers, AI ethics consultants, and prompt engineers are on the rise, but specialized skill sets are necessary.

Many wonder if societies and businesses are prepared to retrain or redistribute a displaced workforce.

In the realm of data privacy, the new currency of the digital economy is personal information. Apps and platforms collect vast amounts of data, training algorithms to predict consumer behavior, drive targeted ads, or refine product recommendations. Yet episodes like the Cambridge Analytica scandal highlight ethical pitfalls. Regulators, notably in the European Union with the General Data Protection Regulation (GDPR), have tried to rein in misuse of data, mandating disclosure and consent. Still, tensions persist about who truly profits from user data, and how to ensure fairness, respect, and benefit-sharing in an AI-driven world.

> *"The fundamental question of our time is how to govern this technology well so it serves humanity's needs rather than undermines them."*
> Brad Smith (president of Microsoft)

CHANGING HUMAN BEHAVIOR AND REQUIRED SKILL SETS

Alongside these sweeping economic changes, AI influences how individuals navigate daily life. Many of us habitually use voice assistants like Alexa, Siri, or Google Assistant for tasks ranging from setting morning alarms to controlling smart home devices. Meanwhile, social media feeds are algorithmically tailored to our perceived preferences, sometimes creating 'filter bubbles' that limit exposure to diverse ideas. Streaming platforms track our viewing or listening habits, recommending new shows or music with impressive accuracy. While these conveniences can save time and effort, they also foster a growing dependence on AI's 'smart suggestions.'

In education, AI-based solutions can personalize learning for students, adapting the difficulty of questions to each learner's progress. Universities worldwide now incorporate AI modules across various disciplines – business, humanities, law – to ensure graduates remain relevant in an automated future. Beyond formal schooling, self-paced online courses from platforms like Coursera or Udemy have become central for reskilling and upskilling.

Still, the relationship between humans and AI is complex. Overreliance on automated systems might diminish critical thinking and problem-solving skills. Some mental health experts voice concerns about potential isolation or social disconnection when large portions of our decisions and interactions pass through algorithmic filters. At the same time, defenders argue that by automating routine tasks, AI frees humanity for higher-order activities like creativity, empathy, strategy, and relationship-building.

> *"AI is probably the most important thing humanity has ever worked on. We need to make sure it complements us, not replaces us."*
> Sundar Pichai (CEO of Alphabet/Google)

EMERGING AI-DRIVEN BUSINESS MODELS

A decade ago, launching a venture that relied heavily on AI might have required specialized infrastructure, deep pools of funding, and extensive academic expertise. That landscape has shifted dramatically. As we said, the rise of open-source libraries such as TensorFlow, PyTorch, and Scikit-learn, combined with the widespread availability of pay-as-you-go cloud services, has democratized AI development. Startups from any corner of the world can prototype and deploy AI solutions without building large in-house data centers.

As a result, early-stage companies can innovate quickly, refining their products based on real-time feedback and user analytics. This dynamic fosters a continuous loop of data collection and model improvement – a hallmark of modern AI-led enterprises. Established corporations are also embracing AI, relying on sophisticated analytics for marketing, logistics, or HR processes. Many have set up internal 'AI centers of excellence' or partner with external data science firms.

Increasingly, providers offer AI as a Service (AIaaS). Rather than coding an algorithm from scratch, a developer can simply tap into a provider's application programming interface (API) for tasks like image recognition, sentiment analysis, or translation. This reduces barriers to entry and helps smaller organizations compete with industry giants. But it also raises questions about data ownership, vendor lock-in, and the resilience of organizations that rely extensively on external AI providers.

Data monetization has become a robust segment in its own right. Retailers mine purchasing trends, user behaviors, and supply chain metrics to create analytics packages. These can be sold or shared with business partners or used internally to improve efficiency. In this paradigm, data is a tradable asset – raising the stakes on privacy, security, and regulatory compliance.

Meanwhile, society expects ethical AI. Growing awareness of algorithmic bias, data security lapses, and misaligned incentives has prompted calls for transparency, fairness, and accountability. Some companies now carry out bias audits or publish annual AI ethics reports. Others consult ethicists to guide product design, ensuring that technology remains inclusive.

> *"Technology should serve humanity. We have a moral responsibility to ensure AI doesn't create more problems than it solves."*
> Tim Cook (CEO of Apple)

THE SYNERGY OF AI AND BLOCKCHAIN IN WEB3

For many, Web3 is the logical evolution of the internet – a world where data ownership is decentralized, user autonomy is enhanced, and communities govern themselves through transparent protocols. Meanwhile, AI thrives on data, gleaning insights from vast swaths of information. Bringing the two together promises intriguing possibilities and unique tensions.

In a traditional AI model, data often resides in centralized servers – usually owned by a major tech corporation. They control access, usage, and distribution. By contrast, blockchain technology aims to distribute record-keeping across a network of participants. Each transaction or data entry is stored cryptographically and immutably, which fosters transparency and trust. The question is: can we combine the power of AI's pattern-finding capabilities with the trustless environment of blockchain?

One concept is 'decentralized intelligence.' Blockchain-based markets or data lakes could let individuals retain ownership of their data. Whenever an AI developer wants to train on that data, a transparent smart contract ensures the rightful owners are compensated. This not only incentivizes data sharing but also promises a more equitable distribution of AI's benefits. In a sense, the data economy becomes more democratized.

Another angle is verifying the authenticity of information used to train AI models. Machine learning is famously prone to the 'garbage in, garbage out' problem. If data is flawed, biased, or tampered with, the resulting models can be misleading. Blockchain's immutability means data sources can be traced, giving AI developers and end-users a reliable record of data provenance. That's particularly relevant in industries like supply chain management, where sensor data about temperature or location is essential for ensuring product quality and integrity.

Additionally, there is discussion of AI oracles for smart contracts. DeFi protocols or decentralized applications often need external input – market prices, weather data, shipping updates – to execute code. An AI oracle can analyze multiple data feeds in real time, filter out noise, and deliver aggregated insights to a smart contract. For instance, a decentralized insurance agreement might hinge on AI's detection of natural disasters, automatically disbursing payouts to farmers if certain climate indicators are met.

Still, challenges arise. Blockchain's default transparency might conflict with AI's hunger for large sets of personal or sensitive information. Zero-knowledge proofs and privacy-preserving computations offer potential solutions, letting AI glean insights from encrypted data without revealing the raw details. Governance is another issue: if a decentralized autonomous organization (DAO) employs AI-driven treasury management, members must ensure the model's decisions align with the community's shared values.

"AI and blockchain can together form new trust layers. Decentralized frameworks ensure a level of transparency AI traditionally lacks, while AI gives blockchains the intelligence to adapt to real-world complexities."
Vitalik Buterin (co-founder of Ethereum)

UNITING AI AND DECENTRALIZED FINANCE

The DeFi movement, built atop blockchain, reimagines financial services for a trustless environment. Instead of banks or brokers, smart contracts handle tasks like lending, borrowing, and yield farming. This system has seen exponential growth since around 2020, with billions of dollars locked into DeFi platforms at any given time. AI's entrance into DeFi could substantially reshape how digital assets are managed, traded, and safeguarded.

Risk management is a prime area. In DeFi lending, borrowers post collateral in the form of cryptocurrency, and if the market plummets, that collateral can be liquidated. AI can spot market anomalies or sudden bursts of volatility far faster than human analysts. By drawing on an ocean of on-chain data (transaction patterns, user behavior) plus external cues (economic indicators, social media sentiment), an AI engine might dynamically adjust collateral requirements or interest rates. This helps avoid mass liquidations and stabilizes lending pools – a benefit to both lenders and borrowers.

"With AI, analyzing billions of transactions on-chain becomes possible at scale. That level of transparency and intelligence could give birth to more robust financial products and safer markets."
Changpeng Zhao (co-founder of Binance)

Yield optimization is another AI-driven avenue. DeFi users often shift tokens between lending platforms or liquidity pools, seeking higher returns (a process called yield farming). However, these opportunities can fluctuate quickly. An AI agent, scanning multiple blockchains and protocols, could autonomously move funds, maximizing returns in real time. Traders already use AI bots for tasks like flash loans or arbitrage, but these systems typically require significant manual fine-tuning. A more advanced AI might learn from continuous experimentation, refining its strategy over time.

Inclusive credit scoring is also on the horizon. Traditional finance relies on established credit histories, leaving billions of people – especially in developing regions – outside the system. In DeFi, a user's on-chain activity might be the primary record: how consistently they repay loans, how frequently they stake or trade, and how well they manage risk. AI models can interpret these patterns to generate a unique 'crypto credit score,' enabling underbanked individuals to access loans without physical collateral or formal credit checks.

Yet the complexities of AI in DeFi are real. Highly sophisticated algorithms could obscure the logic behind certain decisions, conflicting with DeFi's transparency ethos. Data oracles remain a vulnerable point – if they feed incorrect information, even the best AI fails. And if a DAO hands over key financial decisions to AI, accountability becomes blurred. These obstacles are surmountable with careful design, robust governance, and open discussions among communities.

> "We've built trustless lending markets, but AI can help refine them. Imagine dynamic rate adjustments, credit risk analysis that learns with each block, and cross-chain liquidity insights all in real time."
> Stani Kulechov (founder of Aave)

ILLUSTRATIVE AI APPLICATIONS IN THE REAL WORLD

While theoretical debates about AI's ramifications can become abstract, concrete implementations help ground the conversation. In the automotive space, residents of Phoenix and Arizona, see Waymo's self-driving taxis operating on city streets. These vehicles rely on deep neural nets to interpret sensor data – LIDAR, radar, cameras – and navigate roads, with collision rates reportedly lower than average human-driven vehicles. Beyond just self-driving cars, city planners are installing AI-driven signals that adjust traffic lights based on congestion patterns, cutting average commute times and emissions.

In healthcare, a hospital in Tokyo uses AI to recommend tailored cancer treatments, analyzing genetic markers and patient histories to identify targeted therapies with better success rates. Similarly, an AI-based system in London helps detect diabetic retinopathy at an early stage, preventing thousands of cases of blindness each year.

Creative fields are not immune to AI's influence. A graphic designer in Toronto uses generative AI to brainstorm digital art pieces, layering the designer's artistic direction with the AI's uncanny ability to generate multiple variations of color palettes, shapes, and compositions. This collaborative approach speeds up the design process and inspires new aesthetic directions.

Environmental efforts also benefit from AI. In Brazil, reforestation campaigns employ drone fleets controlled by machine learning algorithms, mapping out deforested areas and tracking newly planted saplings over time.

Concurrently, satellite imagery analyzed by AI highlights illegal logging zones, enabling faster action from local authorities.

In finance, the scope is vast. Robo-advisors help everyday consumers manage portfolios automatically, balancing risk according to user preferences. Meanwhile, high-frequency trading – enabled by AI – dominates market volume on stock exchanges, scanning micro-fluctuations and making trades in mere fractions of a second.

Collectively, these use cases illustrate AI's capacity to streamline complex tasks, augment human intelligence, and scale decision-making to levels previously impossible. Yet they also hint at a world in which we must pay constant attention to ethics, governance, and potential unintended consequences.

ETHICAL CONCERNS AND GOVERNANCE

As AI grows more sophisticated, so do moral questions around fairness, transparency, and accountability. One prominent issue is bias: if an AI system is trained on data that underrepresents certain demographics or encodes historical prejudices, it can perpetuate discrimination in lending, hiring, policing, or medical decisions. In response, companies and researchers are developing tools to interpret black-box algorithms. Techniques like LIME (Local Interpretable Model-Agnostic Explanations) or SHAP (SHapley Additive exPlanations) attempt to show the factors that influence AI's predictions, helping auditors and developers detect biased reasoning.

Explainability also matters in high-stakes domains such as healthcare or criminal justice. Deep neural networks are notoriously opaque in their internal logic. If a system recommends a surgical procedure, or a judge relies on risk assessments for parole, it is vital to understand the rationale. Regulators increasingly seek 'right to explanation' policies, forcing organizations to provide comprehensible justifications for automated decisions.

Privacy is another flashpoint. AI relies on large volumes of data – often sensitive, personal, or behavioral. Tools like federated learning have emerged to minimize raw data transfer: models train locally on user devices, only sending aggregated updates to a central server. Homomorphic encryption allows computations on encrypted data without exposing it. Such techniques might ease the tension between AI's appetite for data and individuals' right to privacy.

The specter of technological unemployment looms as well. The World Economic Forum forecasts that AI may displace millions of roles while creating millions more in emerging sectors. Nonetheless, transitions can be painful, especially for workers in industries threatened by automation.

Policymakers and business leaders face pressure to design social safety nets or retraining programs so that entire populations are not left behind.

Finally, AI governance on a global scale remains an open challenge – just like we saw in the previous chapter on compliance when looking at Web3 more broadly. Various governments, from the European Commission to national legislators worldwide, are drafting regulations that define standards for data handling, algorithmic fairness, and accountability. Yet AI transcends borders. International collaboration might be required to ensure consistent safeguards and shared frameworks.

> *"We're in a new era. We need frameworks so that innovation can continue, but people can trust these systems."*
> Bill Gates (co-founder of Microsoft)

VISIONS OF THE FUTURE: AI'S CONTINUING EVOLUTION

Speculation on AI's future unfolds along several lines. One concept is hybrid intelligence, in which humans and AI collaborate seamlessly. Humans bring creativity, emotional intelligence, and moral reasoning; machines provide computational speed, pattern recognition, and memory.

As 5G networks proliferate, connected devices – phones, wearables, smart home sensors – gain the capacity to perform AI computations on-device rather than outsourcing to remote servers. This reduces latency, preserves bandwidth, and can address privacy worries by keeping user data local. Wearables can monitor health in real time; industrial sensors can tweak production lines instantaneously; drones can navigate unfamiliar terrain without round-trip communications.

Another frontier is bridging AI with the metaverse. Virtual spaces, whether for gaming, social engagement, or business collaboration, can become richer when AI procedurally generates environments, dialogues, and events. Non-player characters can exhibit sophisticated behaviors, responding to user actions with near-human adaptability. AI-driven content creation could also help brand experiences in immersive realms.

Some observers debate the path toward AGI (Artificial General Intelligence), machines that can learn and perform across the full range of cognitive tasks humans can tackle. While this remains speculative, large research labs and organizations continue funding advanced research into reasoning, creativity, language understanding, and autonomy. For many experts, the immediate focus should be making current AI systems robust, fair, and transparent rather than rushing headlong toward AGI.

Policy frameworks and social contracts will also evolve, potentially spawning 'AI bills of rights' to protect individual freedoms. Governments and private institutions may create data ownership statutes – guaranteeing that people can track, control, and benefit from the personal data used to train AI systems. Collaborative governance across countries, likely through bodies akin to the OECD or special AI councils, might become the norm to prevent regulatory patchworks.

> *"The future belongs to those who can blend AI with human empathy, creativity, and compassion. The real edge will come from applying AI in ways that deeply serve society."*
> Dr. Kai-Fu Lee (AI investor and author)

REFLECTING ON THE PATH FORWARD

The final takeaway is that AI's journey is far from complete. While it has transformed industries, lifestyles, and data-driven decision-making, the next frontier will likely see AI integrate deeper into decentralized infrastructures. The synergy of AI and blockchain offers a realm of possibilities for more equitable data economies, more secure and adaptive smart contracts, and a new wave of financial inclusion driven by DeFi. The only thing we can say for sure is that the AI of today is the worst it can ever be in the future, at least, from a technical perspective. The leaps forward, often major, will continue to come at us with ever increasing speed.

Yet, caution is warranted. Societies must navigate the ethical, social, and regulatory challenges that inevitably accompany such rapid innovation. Bias in algorithms, potential job displacement, data privacy concerns, and the threat of 'black box' decision-making all demand thoughtful attention. The impetus lies with governments, businesses, and individual citizens to craft a balanced framework ensuring AI is harnessed responsibly.

In Yemi's case, bridging local farming communities to consumers via AI highlights a hopeful trajectory: technology can be a great equalizer, enabling creativity, economic opportunity, and community empowerment. As we scale that story from an urban center in Lagos to a global stage, the opportunities multiply. AI can help expand access to finance, accelerate scientific breakthroughs, overhaul educational systems, and coordinate ecological conservation efforts. Paired with the Web3 vision – decentralization, user agency, and open collaboration – the potential to build a more inclusive and robust digital frontier becomes increasingly tangible.

AI AS A LINCHPIN OF TOMORROW'S INTERNET

The arc of AI's transformation – from an elusive academic goal to a mainstream powerhouse – mirrors the evolution of the internet itself. Where Web1 offered static pages and limited interactivity, Web2 introduced platforms shaped by user-generated content but largely controlled by centralized corporate entities. Web3 proposes a new paradigm of decentralization, ownership, and trustless transactions. AI, with its ability to interpret complex data, automate key functions, and accelerate innovation, stands poised to be the engine driving this next generation of the web.

We see how AI strategies now infuse daily decisions, from personal shopping and entertainment choices to credit allocations and business forecasts. Meanwhile, the responsibilities that come with such power loom just as large. AI should serve humanity's broader goals: equity, sustainability, security, and authentic engagement. When woven into a decentralized architecture, AI can foster transparent, verifiable systems where every participant has a voice in governance, data usage, and incentive distribution.

For readers, the message is both cautionary and optimistic. As society grapples with potential pitfalls that we have discussed above – algorithmic bias, privacy risks, labor displacement – there is an equal (if not greater) measure of promise. Skilled professionals can channel AI to solve pressing global challenges, from healthcare disparities to climate change. Communities can reclaim ownership of their data and shape the rules of engagement. Startups like Yemi's can scale beneficial ideas rapidly, unencumbered by cost barriers or data scarcity.

Ultimately, AI's trajectory will be shaped by the values, ethics, and collective choices we embed into its code and governance frameworks. Perhaps more than any other technology, AI reminds us of our power to create – and our responsibility to use that power wisely. As we stand at the cusp of a new internet era, one characterized by DeFi, tokenized assets, and community-driven platforms, AI emerges as a vital ally in forging a more inclusive and dynamic digital future.

With ongoing collaboration, rigorous self-examination, and innovative engineering, AI can truly elevate humankind. The digital horizon stretches wide, illuminated by the neural networks we've crafted. And as we step forward into the realm of Web3, guided by the strength of blockchain's trustless architecture, we can anticipate an era where intelligence – both artificial and human – intertwines to spark bold solutions for the challenges awaiting us. The path is bright, the momentum is unstoppable, and the revolution of AI and Web3 has only just begun.

ACTIONABLE STEPS : ADOPTING AI AND PREPARING FOR THE AI-DEFI REVOLUTION

As AI and DeFi come together to transform the world, taking practical steps is essential. Yemi's venture – connecting local farmers with urban grocers – shows that even small projects can lead to significant change. In this final segment, we present a clear roadmap of actionable steps inspired by the evolution of AI, the rise of Web3, and the promising future of DeFi.

1. Build a solid foundation in AI literacy

Embrace structured learning: Before deploying AI solutions, ensure you have a robust understanding of key concepts. Start by enrolling in introductory courses on platforms such as Coursera, Udemy, or specialized programs offered by emerging hubs like Roster. These courses should cover:

- **Data fundamentals**: Data cleaning, pre-processing, and the basics of statistical analysis.
- **Machine learning essentials**: Supervised and unsupervised learning, model evaluation, and an introduction to neural networks.
- **Advanced topics**: Deep learning architectures, natural language processing, and reinforcement learning.

Continuous skill development: AI is an ever-evolving field. To stay current:

- **Subscribe to reputable newsletters**: Follow publications and expert blogs to receive timely updates.
- **Engage with thought leaders**: Platforms like LinkedIn provide access to expert opinions, case studies, and emerging trends.
- **Participate in workshops and conferences**: Events not only offer insights into the latest research but also foster networking opportunities with like-minded innovators.

2. Integrate AI into daily workflows

Start small with off-the-shelf tools: Incorporate existing AI solutions into your day-to-day operations to build confidence and generate immediate value:

- **Chatbots for customer service**: Automate repetitive queries to streamline customer engagement.
- **Data analytics**: Use machine learning algorithms to mine sales data for actionable insights and trend detection.
- **Creative assistants**: Employ generative AI tools to support brainstorming sessions or creative projects.

Launch pilot programs: Pilot projects are the proving ground for AI initiatives. Consider these starting points:

- **Recommendation engines**: Develop a small-scale system to personalize product suggestions for e-commerce platforms.
- **Helpdesk automation**: Implement AI-driven systems to handle common customer inquiries, reducing wait times and operational costs.

- **Quality assurance in production**: Use computer vision for real-time monitoring of manufacturing defects or process inefficiencies.

Pilot programs offer low-risk environments to test hypotheses, learn from real-world data, and iteratively improve solutions before scaling them across larger operations.

3. Scale and optimize AI initiatives

Transition from pilots to enterprise-wide adoption: Once initial pilots have proven effective, carefully expand AI integration by the following ways:

- **Incremental rollouts**: Gradually extend successful projects across departments or geographical regions.
- **Iterative feedback mechanisms**: Establish regular reviews with cross-functional teams to fine-tune AI systems based on user feedback and evolving business requirements.
- **Collaboration between business and technical teams**: Foster environments where data scientists, engineers, and business leaders jointly identify opportunities for scaling and innovation.

Enhance data infrastructure: A scalable AI strategy depends on robust data pipelines:

- **Invest in data quality**: Ensure that data from internal sources, customer interactions, and even third-party platforms is clean, well-structured, and representative.
- **Adopt cloud-based platforms**: Utilize scalable cloud services to manage data storage, processing, and computational demands.

4. Establish robust governance and ethical frameworks

Develop an internal ethics framework: As AI systems become integral to decision-making, embedding ethics into every stage of development is crucial:

- **Create guiding principles**: Establish a set of ethical guidelines that cover fairness, transparency, and accountability. This might include forming an internal ethics committee to oversee AI projects.
- **Regular audits**: Implement tools like LIME or SHAP to interpret AI decision-making processes and routinely audit systems for biases or ethical breaches.

Ensure regulatory compliance: In an era of increasing data privacy concerns you must be ahead of the curve on key issues, such as:

- **Stay informed on global regulations**: Follow updates on policies such as the GDPR and similar frameworks worldwide.

- **Implement privacy-preserving techniques**: Use federated learning and homomorphic encryption to protect sensitive data while still extracting valuable insights.

5. Explore the synergy between AI and DeFi

Understand DeFi basics: Familiarize yourself with the DeFi landscape by studying key protocols like Aave, MakerDAO, and Uniswap. This knowledge lays the groundwork for integrating AI into a trustless financial ecosystem.
  Leverage specialized platforms: To combine AI with blockchain:

- **Engage with platforms like Ocean Protocol and Singularity NET**: These ecosystems facilitate secure data sharing and enable AI applications in decentralized settings.
- **Experiment with AI-driven oracles**: Develop or adopt systems that bridge on-chain data with off-chain analytics, ensuring that smart contracts receive real-time, accurate information.

Implement practical use cases in DeFi: Start with low-risk applications that demonstrate tangible benefits:

- **Risk management in lending**: Use AI to analyze market trends, adjust collateral ratios dynamically, and anticipate market shifts before they trigger mass liquidations.
- **Yield optimization**: Deploy AI agents that monitor multiple blockchains for yield farming opportunities, autonomously shifting assets to maximize returns.
- **Inclusive credit scoring**: Build AI models to assess on-chain behavior, offering a new credit score for individuals without traditional financial histories, thereby promoting financial inclusion.

6. Foster a collaborative and adaptive ecosystem

Cultivate internal and external partnerships: The AI-DeFi revolution is best tackled through collaboration, and so you may consider:

- **Cross-functional teams**: Bring together experts from AI, finance, legal, and regulatory backgrounds to create well-rounded solutions.
- **Industry partnerships**: Forge alliances with academic institutions, technology incubators, and blockchain communities to tap into cutting-edge research and development.

Build a culture of knowledge sharing: Encourage continuous learning and internal dialogue:

- **Host seminars and hackathons**: Create forums for sharing successes, challenges, and innovative ideas.

- **Develop an internal resource hub**: Maintain a repository of best practices, case studies, and technical guides that employees can reference.

Maintain agility and adaptability: The landscape of AI and DeFi is dynamic. To stay ahead:

- **Schedule regular strategy reviews**: Periodically assess the performance of AI initiatives and adjust strategies based on market trends and technological advancements.
- **Promote lifelong learning**: Ensure that teams have access to ongoing training, industry conferences, and online learning resources to continuously update their skills.

7. Envision a responsible and inclusive future

Align technology with social impact: Beyond technical excellence, consider the broader societal implications of your AI-DeFi initiatives:

- **Empower local communities**: Just as Yemi's project transformed local agriculture in Lagos, seek ways to use technology to address regional challenges and create equitable opportunities.
- **Promote transparency and accountability**: Embed systems that allow stakeholders to understand, question, and influence AI-driven decisions – ensuring technology remains a tool for empowerment rather than exclusion.

Prepare for the next frontier: Recognize that the AI of today is the stepping stone for the breakthroughs of tomorrow, just take a moment to pause and think, what can you do today that you couldn't do yesterday (thanks to technology it is quite a lot):

- **Invest in research and development**: Allocate resources to explore emerging areas like hybrid intelligence, on-device AI computations, and the integration of AI within the metaverse.
- **Plan for scalability and resilience**: Design systems that not only perform under current conditions but can also adapt to future challenges, regulatory shifts, and technological evolutions.

Conclusion: Charting a bold course in the AI-DeFi era

The convergence of AI and DeFi represents a transformative opportunity for all companies and individuals alike – a chance to create systems that are more efficient, transparent, and inclusive. From building a foundational understanding of AI to deploying sophisticated, scaled projects and embedding robust governance frameworks, the steps outlined in this chapter serve as a comprehensive guide for navigating this complex, rapidly evolving landscape.

For entrepreneurs, corporate decision-makers, and curious individuals alike, the future is not just a destination but a journey of continuous learning, adaptation, and collaboration. By taking proactive, measured steps today, you can harness the dual power of AI and DeFi to not only drive business success but also contribute to a more equitable and sustainable digital future.

The path forward demands courage, ingenuity, and a steadfast commitment to ethical innovation. With each strategic initiative and every thoughtful pilot project, we edge closer to a world where AI and DeFi work in harmony – unlocking opportunities, reducing inefficiencies, and ultimately, transforming the way we interact with technology and one another.

As you embark on this journey, remember that the revolution of AI and Web3 is already underway. The insights and strategies presented here are designed to empower you to seize the moment, mitigate risks, and lead the charge toward a future where technology serves humanity's greatest aspirations.

Chapter 11

The Web3 imperative

Embracing a decentralized future today

As we come to the end of this journey through the world of Web3, it's important to take a step back and reflect on everything we've learned. What have we uncovered? What is the promise of Web3, and how does it truly reshape not just business, but society itself? These are the questions we must all consider as we embark on the next chapter in the digital world.

Web3 is not just a new version of the internet. It is not merely about the latest technological trends like **blockchain, smart contracts,** or **NFTs**. It is, at its core, a **revolution** – one that challenges long-held beliefs about ownership, control, and the distribution of value. In a world dominated by centralized systems, Web3 offers a blueprint for a more **equitable, transparent,** and **empowering** digital economy. It's a space where power is redistributed from a few central authorities to the people who participate in the digital world – the creators, innovators, consumers, and communities who drive the economy forward.

In this final chapter, we'll take a look back at the transformative journey we've explored throughout the book. From the **foundational technologies** of blockchain to the **societal shifts** Web3 brings, we've discussed the incredible opportunities, as well as the challenges, that Web3 presents. But more importantly, we'll look ahead – toward the future. What does that future look like? And how can we actively contribute to making it a reality? Together, we'll reflect on the transformative potential of Web3 and challenge ourselves to think about how we can shape the future of the digital world.

THE PROMISE OF WEB3: A NEW DIGITAL PARADIGM

Web3 is not just a technological evolution; it's a **paradigm shift**. Throughout this book, we've seen how Web3 enables **decentralized systems** – systems that bypass the traditional intermediaries that have long controlled financial transactions, content creation, and even personal data. At the core of Web3 lies blockchain, the technology that underpins these decentralized systems. Blockchain allows for **trustless transactions** – transactions that do

DOI: 10.1201/9781003616504-11

not require a third party to verify the exchange, as everything is verified by the network itself. This simple yet powerful shift is what makes Web3 so revolutionary.

Web3 represents a move away from centralized platforms – be it social media giants, e-commerce platforms, or financial institutions – to a world where individuals regain control. Web3 offers users **ownership** over their data, their identities, and even the content they create. In Web3, there is no central entity that dictates the rules; instead, decision-making is **distributed** across a **decentralized network**, often through **smart contracts** and **governance tokens**. This democratization of power is the essence of what makes Web3 so different from what we have today.

The most exciting aspect of Web3, perhaps, is its potential to **transform industries** in profound ways. We've already seen the rise of **DeFi** (decentralized finance), a new financial system that enables individuals to lend, borrow, and trade assets without needing a central bank or intermediary. This shift allows for **peer-to-peer finance**, creating opportunities for people who are excluded from traditional financial systems. It empowers individuals in **emerging markets**, gives them access to financial services, and allows them to participate in the global economy.

The rise of **NFTs** (non-fungible tokens) and the **Metaverse** are also profound transformations that are happening right now. NFTs allow creators to tokenize their digital content – whether it's art, music, or even tweets – and sell it directly to their audiences. This new model of ownership and monetization gives creators the ability to **retain control** over their intellectual property and, most importantly, receive fair compensation for their work. The Metaverse, an immersive virtual space built on Web3 principles, is bringing the digital world into an entirely new dimension, offering opportunities for **virtual economies** and experiences that go beyond the limits of the physical world.

In these ways, Web3 allows us to **reimagine** entire sectors – finance, entertainment, governance, and beyond – creating new ways to build and exchange value that are **more inclusive, transparent,** and **efficient.**

WEB3'S SOCIETAL IMPACT: THE POWER OF EMPOWERMENT AND OWNERSHIP

One of the most compelling features of Web3 is the **empowerment** it offers to individuals. In Web2, our digital identities, data, and transactions were controlled by a handful of centralized platforms – companies like Alphabet, Meta, and Amazon. These entities accumulated vast amounts of personal information, which they used to profit from advertising and user data. This model often led to a sense of **disempowerment** for users, who had little say in how their information was used or monetized.

Web3, in contrast, is about **returning control** to the individual. It enables **self-sovereign identity** – the ability to control your digital identity without relying on central authorities. Blockchain enables users to take control of their own **data**, ensuring that it cannot be exploited by corporations or governments. In this new world, individuals – not corporations – hold the keys to their own digital lives. This empowerment extends to **monetary value** as well. With Web3, users can **create, own,** and **monetize** their content in ways that were not possible before.

For businesses, this means moving away from models that exploit user data and instead embracing a **community-driven approach** that values **transparency** and **trust**. Web3 businesses will be built around the principles of **collaboration,** where users are not just consumers, but active participants in the governance and growth of the platform. In the world of Web3, companies must think differently about their relationship with their users. Instead of extracting value from users, businesses must **create value** for their communities.

Perhaps the most important lesson of Web3 is the **redistribution of power.** We've spent decades living in a world where power, wealth, and access were concentrated in the hands of a few corporations. In Web3, the playing field is being leveled. This is an opportunity to build a world where digital and financial systems are open to **everyone,** where opportunities for success and wealth creation are available to all, not just the privileged few.

BUSINESS IN THE WEB3 ERA: CREATING MORE ETHICAL, INCLUSIVE MODELS

As we discussed throughout the book, **businesses** in the Web3 space have a tremendous opportunity to build more **ethical** and **sustainable** models. Web3 offers the tools to create transparent, decentralized businesses that prioritize **community** and **collaboration** over profit-maximization and hierarchical structures. This is a chance to shift the mindset from traditional business practices to a more responsible, inclusive, and value-driven approach.

The possibilities for business in Web3 are vast, from **NFT marketplaces** to **DeFi protocols** to **dApps** that can revolutionize how we interact with everything from content to financial services. But with these opportunities comes a need for responsibility. Businesses that operate in Web3 must do so with a focus on **ethics** – ensuring that their platforms are accessible to everyone, that they prioritize **privacy,** and that they are governed in ways that reflect the values of decentralization and transparency. Web3 businesses have the opportunity to break free from the **monolithic corporate structures** of Web2 and build organizations that are rooted in **collective ownership** and **shared success.**

The success of Web3 will not just be measured by financial outcomes; it will be measured by its **social impact** – the degree to which it empowers individuals, supports creators, and fosters inclusivity in digital and economic systems. The businesses of tomorrow will be those that not only profit but **serve the greater good**. In this new world, the goals of business are no longer at odds with the welfare of the people and the planet; instead, the two are inextricably linked.

OVERCOMING CHALLENGES: SECURITY, REGULATION, AND SCALABILITY

As with any revolution, there are challenges to overcome. The transition from centralized systems to a decentralized Web3 world will not be without its hurdles. These challenges are primarily technical, legal, and societal.

Security is one of the most pressing issues facing Web3. While blockchain itself is inherently secure, the platforms and applications built on top of it – whether they're DeFi protocols or NFT marketplaces – are often vulnerable to hacks, exploits, and fraud. As more people engage with Web3 technologies, the need for robust security measures becomes even more critical. Platforms must invest in **cybersecurity** and **audit** their systems rigorously to ensure the safety of user assets.

Regulation is another challenge that looms over Web3. Governments worldwide are still figuring out how to regulate decentralized technologies. The lack of clarity around regulation has created uncertainty for businesses and individuals alike. However, this is an opportunity for the Web3 community to engage with regulators and shape the future of governance in the space. Clear and thoughtful regulation can foster innovation while protecting consumers, and it's essential for creating a safe and sustainable Web3 ecosystem.

Finally, **scalability** remains a challenge. As Web3 grows, it must be able to handle an increasing volume of transactions, users, and data. The blockchain networks that power Web3 technologies must evolve to become faster, more efficient, and capable of supporting global demand. This is an area of intense focus, and as we see the rise of **Layer 2 solutions** and more **efficient consensus mechanisms**, the scalability challenges facing Web3 will continue to be addressed.

A VISION FOR THE FUTURE: THE WEB3 REVOLUTION IS NOW

Looking ahead, the future of Web3 is incredibly bright. The digital economy is rapidly shifting toward decentralization, and the possibilities it presents are endless. From **virtual economies** to **decentralized finance**, from **self-sovereign identities** to **transparent governance**, Web3 is paving the way for a **more open, more inclusive**, and **more equitable** world.

As we've seen, **decentralized systems** allow for more direct and fair interactions between individuals, communities, and businesses. In this new world, everyone has the opportunity to own, create, and contribute. As we move into this new era, we must remember that the true potential of Web3 lies not just in the technologies themselves, but in how we use them to create a better, more just world.

The Web3 revolution is **already happening**, and it's up to us – **creators, entrepreneurs, business leaders, consumers,** and **advocates** – to make sure that this new digital world is one that serves everyone. We must not simply watch from the sidelines but actively engage in shaping this future. Together, we can build systems that empower individuals, elevate communities, and make the digital world more **ethical, transparent,** and **inclusive** than ever before.

A CALL TO ACTION: SEIZE THE OPPORTUNITY

The opportunity to build a more equitable, transparent, and decentralized world is here. Web3 is not just about technology – it is about people. It's about reshaping the way we do business, how we create and consume, and how we govern ourselves. **This revolution is now** – and you, as an entrepreneur, creator, or advocate, have the chance to shape the future of the digital world.

Don't wait for others to make this change. **Lead the charge.** Whether you're building a Web3 project, creating content, developing new decentralized applications, or simply exploring the space, your involvement is crucial. We need more **innovators,** more **dreamers,** and more **builders** to help realize the full potential of Web3. The digital future is in our hands – let's embrace it and build something great, something that works for everyone.

The future is decentralized. The future is Web3. The future is in **your hands.** Let's build it together.

"Welcome to Web3. Your adventure begins now."

Index

in decentralized web of interoperable
applications, 21–22
developing a roadmap for, 123
focus on self-custody and digital
ownership, 15–16
future of, 41, 173–174
future of work and, 22
how to align strategy and culture for,
134–137
how to get started with, 23–24
identifying new technology
opportunities with, 122–123
incentive structure of, 13–14
integration with other emerging
technologies, 22
key pillars in strategy for, 129–132
long term horizons in, 130–131
magic questions for, 131–132
navigating compliance issues with,
138–148
overcoming barriers to adoption of,
20–21, 173
partnerships and, 129–130
promise of, 2, 170–171
real-world examples of, 10–14, 18–20
self-custody and ownership economy
with, 8–10

societal impact of, 171–172
staying agile with, 123–124
strategy and culture for, 133–134
SWOT analysis for, 124–132, 125
testing and learning with, 129
transforming industries, 22–23
understanding users of, 122
web strategy, development of, 121–124
weighted voting, 85
whitepapers, decentralized autonomous
organizations (DAOs), 84
Winkelmann, Mike, 65
Winklevoss, Tyler, 46

X
X (formerly Twitter), 7, 16, 70

Y
YellowHeart, 73
yield farming and staking, 58, 61
YouTube, 7

Z
zero-knowledge proofs (ZKPs),
59–60, 158
Zhao, Changpeng, 159
Zuckerberg, Mark, 114

For Product Safety Concerns and Information please contact our EU
representative GPSR@taylorandfrancis.com
Taylor & Francis Verlag GmbH, Kaufingerstraße 24, 80331 München, Germany

www.ingramcontent.com/pod-product-compliance
Lightning Source LLC
Chambersburg PA
CBHW070718220326
41598CB00024BA/3220